国家双语教学示范课程配套教材

Agroecology

农业生态学

（第二版）

王松良　〔加〕C.D. 考德威尔（Claude D. Caldwell）　主编

科学出版社

北京

内 容 简 介

本书是"国家双语教学示范课程"配套教材,也是福建农林大学与加拿大戴尔豪斯大学(Dalhousie University)本科教育合作项目(2012 年至今)的成果之一。

本书紧紧围绕"农业是把太阳光转变成人们健康、幸福生活的产业"这一在世界农业面临全面生态化转型背景下对"农业"的本质过程和根本目标的理解,把"农业生态学"定义为地球演化进入"人类世"的农业系统学科,它既是指导农业生态转型的一种科学,也是推进农业生态转型的一种实践,更是实现全球可持续的食品体系的一种社会运动。为此,本书主要介绍农业生态学的学科内涵、农业生态系统的基本过程及人为问题,以及可持续农业生产系统的构建与管理等内容。

本书各章编写采用最新的基于学生学习产出评价的课程教学方法,视角独特,图文并茂,适合作为高等农林院校农业生态学课程的双语教学用书,也可用于农业院校和科研院所研究生考试的参考书,以及作为全国农业生态化转型背景下的生态农业管理和实践者的参考书。

图书在版编目(CIP)数据

农业生态学 = Agroecology:汉、英 / 王松良,(加)C.D.考德威尔(Claude D. Caldwell)主编. —2 版. —北京:科学出版社,2021.6

国家双语教学示范课程配套教材

ISBN 978-7-03-068415-8

Ⅰ. ①农… Ⅱ. ①王… ②C… Ⅲ. ①农业生态学-双语教学-高等学校-教材-汉、英 Ⅳ. ①S181

中国版本图书馆 CIP 数据核字(2021)第 048990 号

责任编辑:丛 楠 / 责任校对:严 娜
责任印制:张 伟 / 封面设计:迷底书装

科 学 出 版 社 出版
北京东黄城根北街 16 号
邮政编码:100717
http://www.sciencep.com
北京凌奇印刷有限责任公司 印刷
科学出版社发行 各地新华书店经销
*

2012 年 1 月第 一 版 开本:787×1092 1/16
2021 年 6 月第 二 版 印张:20 1/4
2023 年 1 月第三次印刷 字数:486 000

定价:69.80 元
(如有印装质量问题,我社负责调换)

编委会名单

主　编

S. Wang

College of Agriculture, Fujian Agriculture and Forestry University, Fuzhou, China

C. D. Caldwell

Department of Plant, Food, and Environmental Sciences, Dalhousie University, Faculty of Agriculture, Truro, Nova Scotia, Canada

其他编者

D. H. Lynch

Dalhousie University, Truro, Nova Scotia, Canada

S. Smukler

University of British Columbia, Vancouver, BC, Canada

T. Tennessen

Department of Plant, Food, and Environmental Sciences, Dalhousie University, Faculty of Agriculture, Truro, Nova Scotia, Canada

N. Mclean

Dalhousie University, Truro, Nova Scotia, Canada

K. Pruski

Dalhousie University, Truro, Nova Scotia, Canada

J. Duston

Dalhousie University, Truro, Nova Scotia, Canada

Q. Liu

Department of Fisheries and Oceans, Government of Canada, Vancouver, British Columbia, Canada

T. McKenzie
Dalhousie University, Truro, Nova Scotia, Canada

G. C. Cutler
Dalhousie University, Truro, Nova Scotia, Canada

A. M. Hammermeister
Organic Agriculture Centre of Canada, Truro, Nova Scotia, Canada

E. K. Yiridoe
Dalhousie University, Truro, Nova Scotia, Canada

Z. Si
Balsillie School of International Affairs, Wilfrid Laurier University, Waterloo, Ontario, Canada

S. Scott
Department of Geography and Environmental Management, University of Waterloo, Waterloo, Ontario, Canada

Foreword

The past half-century has seen substantial increases in global food production. World population has risen 2.5 fold since 1960 and yet per-capita food production has grown by 50% over the same period. At the same time, evidence shows that agriculture is the single largest cause of biodiversity loss, greenhouse gas emissions, consumptive use of freshwater, loading of nutrients into the biosphere (nitrogen and phosphorus), and a major cause of pollution due to pesticides. This is manifested in soil erosion and degradation, pollution of rivers and seas, depletion of aquifers, and climate forcing.

This desire for agricultural systems to produce sufficient and nutritious food without environmental harm, and going further to produce positive contributions to natural, social, and human capital, has been reflected in calls for a wide range of different types of more sustainable agriculture. The dominant paradigm for agricultural development centres on intensification (productivity enhancement) without integrating sustainability. When the environment is considered, the conventional focus has been on reducing negative impacts rather than exploring synergies between intensification and sustainability. This suggests the need for a new phase based on agroecology.

As agroecosystems are considerably more simplified than natural ecosystems, some natural properties need to be designed back into systems to decrease losses and improve efficiency. For example, loss of biological diversity results in the loss of some ecosystem services, such as pest and disease control. For sustainability, biological diversity needs to be increased to re-create natural control and regulation functions and to manage pests and diseases rather than seeking to eliminate them. Modern agricultural systems have come to rely on synthetic nutrient inputs requiring high inputs of energy, usually from fossil fuels. These nutrients are often used inefficiently and result in losses in water and air as nitrate, nitrous oxide, or ammonia. To meet principles of sustainability, such nutrient losses need to be reduced to a minimum, recycling and feedback mechanisms introduced and strengthened, and nutrients diverted for capital accumulation. Mature ecosystems are now known to be in a state of dynamic equilibrium, forever moving, shifting, and changing. It is this property that buffers against large shocks and stresses. Modern agroecosystems, in prioritizing simplification, rigid control, and stability, have weak resilience.

It is increasingly clear that systems of agricultural management now need fresh redesign if they are to sustain beneficial outc

omes over long periods of time across differing ecological, economic, social, and political landscapes. Redesign is a social and institutional challenge, as landscape-scale changes are needed for positive contributions to biodiversity, water quantity and quality, pest management, and climate change mitigation. We have suggested there are non-linear stages: efficiency, substitution, and redesign.

Efficiency aims to make better use of on-farm and imported resources within existing farm configurations. Many agricultural systems are wasteful, and so on-farm efficiency gains can arise from better management to reduce use, precision targeting of fertilizer, pesticide, and water to cause less damage to natural capital and human health.

Substitution focuses on the replacement of technologies and practices with more sustainable forms. Forms of substitution include the release of biological control agents to substitute for agrochemical inputs, replacing the use of soil by hydroponics, and no-tillage systems that use new forms of direct seeding and weed management to replace inversion ploughing.

Redesign is the stage fundamental for achieving sustainability at geographic scale. Redesign of agroecosystems and landscapes is necessary to harness ecological processes such as predation, parasitism, allelopathy, herbivory, nitrogen fixation, and pollination. While efficiency and substitution tend to be incremental within current production systems, redesign should be the most transformative, often resulting in fundamental changes to system components and configurations.

Many of the recent successes in sustainable agriculture derive from agroecology, the application of ecological principles to the management of both farmed and non-farmed components of agricultural systems. Farming that harnesses ecology by design may thus be able to use fewer or no artificial inputs or compounds. A key focus of such agroecosystem management approaches is increased reliance on knowledge and management (or design), complementing and reducing the use of technological inputs. In 2011, the UN's Food and Agriculture Organization (FAO) also called for a paradigm shift in agriculture, towards sustainable, ecosystem-based production. FAO's latest farming model, *Save and Grow*, aims further to address today's intersecting challenges: raising crop productivity and ensuring food and nutrition security for all while reducing agriculture's demands on natural resources, its negative impacts on the environment and its effects on climate change.

New approaches as exemplified in this book recognize that food security will depend as much on environmental sustainability as it will on raising crop productivity. These seek to achieve both objectives by using improved varieties, drawing on ecosystem services (such as nutrient cycling, biological nitrogen fixation, and pest predation), and minimizing the use of farming practices and technologies that degrade the environment, deplete natural resources, add momentum to climate change, and harm human health.

There is also overwhelming evidence from the field and reported in the published literature that collective management of resources helps redesign and increases net system productivity. The term sustainable suggests an incorporation of the need for improvement (e.g. to well-being, food production, natural capital) and thus requires the need to change the way we as individuals think about and come to know about the world. Sustainability is a form of progressive and gradual change that requires changes in behaviours and practices as well as internal changes to mindsets. It could be that novel forms of social capital can open up science to innovation, particularly where problems are complex and solutions unknown, and where the values of all actors are salient.

Jules Pretty, University of Essex, Colchester, UK

Jules Pretty is Professor of Environment and Society at the University of Essex and the author of *Regenerating Agriculture* (1995), *Agri-Culture* (2002), *The Earth Only Endures* (2008), *The Edge of Extinction* (2014), and *The East Country* (2017).

Preface

This textbook is one of the outcomes of the NSAC-FAFU 2+2 program, an international undergraduate cooperative education program, initiated in 2003 between Fujian Agriculture and Forestry University (FAFU) in China and Nova Scotia Agricultural College (NSAC, which became the Faculty of Agriculture of Dalhousie University in 2012) in Canada. Under this framework, Dr. Claude Caldwell of NSAC introduced the course of Agroecology to FAFU in collaboration with Dr Wang Songliang. The course is currently available for the undergraduate students majoring in Horticulture and Agricultural Resources and Environment at FAFU. In 2007, the course was certified with the name of "Provincial Elite Course" and subsequently was named a model for bilingual education in China in 2009. This course provides an excellent opportunity for the students to become familiar with western teaching and learning styles while using English textbooks and referencing system. This experience has been a key part of the success of many of our new global citizens.

Therefore, the stimulus for this textbook was more than a decade of collaborative, interdisciplinary, and cross-cultural teaching of the discipline of agroecology. Our understanding of both theory and practice has evolved as the two of us taught, discussed, argued, published, shared graduate students. It was a learning experience for all involved. Most satisfying has been watching students take the theory and put it in practice; in return, those students have challenged us to stretch ourselves beyond the classroom and to make more of a difference in the world.

This book is an effort to capture some of what has occurred in our classroom, with the hope that it will stimulate other teachers and students to discover, understand, and apply the principles of agroecology to the benefit of people and the Earth.

Our humanity is now in an Anthropocene Era, an era in which humans have become geophysical forces having the same level of impact as asteroids and volcanoes had in the past. Many activities of human are not sustainable; therefore, we should recognize and learn from our past mistakes and successes. This involves re-evaluating our past behaviour, considering new pathways to sustainability, educating ourselves, and finding opportunities from crises.

In general, the human population faces four interwoven, critical challenges in this Anthropocene Era:

1. *Increasing demand for food.* By 2050, the world population will reach 9 billion, a 35% increase over today's population. This growth is exacerbated by changes in consumption trends. Worldwide, including the so-called developing nations, there is an increasing demand for meat and animal protein. Increasing numbers multiplied by increasing consumption per person means our food production needs to double by 2050 to meet that demand.

2. *Weather chaos due to climate change.* The Anthropocene is characterized by fossil energy use, which has caused an imbalance in a few important gases in our atmosphere, especially CO_2, CH_4, and N_2O. These greenhouse gases (GHGs) are leading to global climate change with the potential for disaster in the future if we fail to take collective action to decrease GHG emission.

3. *Decreasing quality and quantity of freshwater.* Even though our living planet is described as the water planet, the freshwater (potable and usable water) is extremely limited, accounting for less than 0.02% of the total amount of water on earth. Recent human activity, much of it aligned to agricultural production, has significantly deteriorated water quality and quantity. This has led to potable water emergencies in 2/3 of middle- to large-sized cities of the world and threatens global food security.

4. *Human livelihood disruption.* Another explicit characteristic of the Anthropocene era is the shift as societies focus towards business and capital. Development and financial return take priority over human livelihood. This tends to degrade the natural ecosystem and also erodes humanity's potential to achieve global peace and harmony.

A positive future is possible but successful strategies are needed to address all 4 of these interwoven challenges. All of the above are significantly negatively affected by our current global agriculture model. Hence, a sustainable agricultural transformation and food system restructuring strategy, properly coordinated across the world, could be the catalyst for success in all the 4 areas. Using the principles and practices of agroecology, our food system can be transformed:

1. We need a new truly *"green" Green Revolution* in agriculture that can quickly respond to the requirement for doubling of available food by 2050.

2. This must be a *climate-smart* food production strategy that can increase available food while lowering GHG emissions.

3. Changes will include a *water-adapted* agricultural model that can produce food with extreme efficiency of water use and greatly decrease the degradation of our freshwater resource.

4. Another key will be an *ecology-oriented value chain* development that can enhance productivity, decrease waste, and provide meaningful employment.

As an initial step, in this textbook we introduce a new broader, more inclusive understanding of agriculture with a new definition: *Agriculture is the science, art, politics, and sociology of changing sunlight into healthy happy people.* This definition is the core spirit of this *Introduction to Agroecology* textbook.

Using this umbrella definition of agriculture, agroecology becomes an interdisciplinary bridge between ecology and agricultural sciences, rural sociology, and economics. Agroecology therefore becomes simultaneously a science, a practice, and a social movement; it provides the core guidelines for food system transformation worldwide. It is our hope that the teaching of agroecology can provide pedagogical change from reductionist to systematic thinking in agriculture and foodrelated colleges and universities.

As Albert Einstein once said "*we cannot solve our problems with the same thinking we used to create them*", likewise, we cannot employ the obsolete energy and water intensive, corporate model to build the above-mentioned new strategic agriculture. Instead, we urgently need a transformation in agriculture from a mechanistic food and feed based production mindset to one that *prioritizes the interactive good health of humans, animals, and the Earth.*

The major barriers to success and survival in the Anthropocene Era are gaps in shared knowledge and a lack of global co-operation. In a time when greater international consensus and action is needed, the game of zero-sum nationalistic politics threatens to scuttle effective action. The aim of this textbook is to stimulate informed coordinated action through educating the next generation of agroecologists.

Note to Instructors

The intended audience for *Introduction to Agroecology* is first-year undergraduate university students; it may also be of use to higher-level undergraduates with an interest in agriculture and ecology. It is an introductory textbook in applied ecology with particular reference to agricultural systems. The level of science and general knowledge reflects this target group.

The text is divided into five parts with 22 chapters in all. The first two parts "Context of Agroecology" and "Basics of Agroecosystems" provide a firm basis for the deeper study of agriculture from an ecological standpoint.

Part III, "Digging Deeper into Agroecosystems", explores the related issues of hunger, wastes, climate change, and biodiversity. It is suggested that students study these three sections before proceeding to Section 4 or 5.

Part IV, "Application of Agroecosystem Concepts", exposes the student to some of the details around agricultural production and challenges the student to use the concepts and ideas from the first three sections to critically evaluate such production systems.

Part V, "Agroecosystem Management", closes the circle by looking at global solutions and opportunities from both a scientific and social economic standpoint. It is hoped the student will find these last four chapters stimulating and positive.

The information in the chapters is not intended to be a comprehensive review of literature; rather the intent is to give stimulating data and analysis of the issue. If the textbook is used for a one-term course, each of the chapters may serve as preliminary reading for 22 individual topic sessions. Each chapter is headed with student learning objectives for that topic, to guide the student in their study. Obviously, individual instructors may wish to change or supplement this list of learning objectives.

The 3 principal instructors of the Agroecology at FAFU, from left to right, Claude Caldwell, Songliang Wang, Shannon Kilyanek

Songliang Wang, Fuzhou, PR China

C. D. Caldwell, Truro, NS, Canada

Acknowledgements

The authors/co-editors are grateful to the Canadian and Chinese governments for their long-term support of the 2+2 Sino Canadian educational program and, specifically, the development since 2004 of cooperative, interdisciplinary, collaborative teaching of agroecology. We also gratefully acknowledge support for Songliang Wang's scholarly visits to NSAC and UBC during 2005–2006 and 2018–2019, respectively.

Special recognition is given to the significant contribution of Ms Shannon Kilyanek, whose passion for teaching and willingness to challenge students across cultures helped develop and mature both the 2+2 program and the agroecology course. Her input has been invaluable.

Our gratitude especially goes to the chapter contributing authors from University of British Columbia, Dalhousie University, the University of Waterloo, the Organic Agriculture Centre of Canada, and the Department of Fisheries and Oceans for Canada.

We would like to thank Ms. Raagai Priya Chandrasekaran and Ms. Zhu Yu, Springer editors in collaboration with China's Science Press, and Ms Cong Nan, the editor responsible for our first 2012 Agroecology edition and this second edition in China's Science Press.

The production of this textbook was supported by the grant of Designing the Payment for Agroecosystem services in Fujian Province [grant number: FJ2018B070], funded by Fujian League of Social Science Foundation; and the grant of Agroecology Knowledge sprending (graut number: 2020R0138), funded by Fujian Provincial Government; and the grant of Textbook Publication Fund of Fujian Agriculture and Forestry University (FAFu) (grant number: 111971702).

前　言

　　本教材是福建农林大学（FAFU）与加拿大新斯科舍农学院（NSAC）合作的 2+2 教育本科项目的成果之一。该项目于 2003 年秋季对合办的园艺、农业资源与环境专业招生，2004年春季，项目创建者之一的 Claude Caldwell 从 NSAC 引进农业生态学课程，与王松良在 FAFU 对上述 2 个合作专业新生开课，2007 年该课程被评为福建省精品课程，2009 年被评为国家双语教学示范课程，2012 年由科学出版社出版本教材的第一版。同年，NSAC 合并到戴尔豪斯大学（Dalhousie University），项目相应升级为 FAFU-Dalhousie 2+2 教育本科项目，农业生态学合作教学继续推进和提升，乃至于有本书的诞生。

　　农业生态学是农业（学）与生态学的融合，它应用生态学的观点和方法研究整体农业系统，培养学生（对农业）的系统视野。作为中加合作办学项目的引进课程，本课程更是为项目学生熟悉加拿大大学教与学风格提供平台，培养学生（对农业）的国际化视野。基于此，本教材第二版的两位主编者努力邀请中外 10 多位农业生态学领域的顶级专家参与编写本版教材。全书分为五部分，共 22 章。其中，第一部分"Context of Agroecology"的 2 章系统探讨人类世（Anthropocene）背景下农业的可持续发展需要的整体学科与系统辨识；第二部分"Basics of Agroecosystems"的 4 章深入探究农业生态系统的结构和功能；第三部分"Digging Deeper into Agroecosystems"的 4 章深化讨论农业生态系统的各个层次问题及其解决思路；第四部分"Application of Agroecosystem Concepts"的 8 章旨在培养学生应用生态系统的思维、概念和方法与分析和评价分类的农业生产体系，形成各自的可持续发展方向；第五部分"Agroecosystem Management"的 4 章分别从生物防治、有机耕作、系统管理等三个国际通用的农业生态系统管理实践出发，最终结束于农业生态学在实现联合国的"可持续发展目标（Sustainable Development Goals, SDGs）"的作用的阐述上。

　　本教材是基于中加 2+2 本科合作项目（专业）一年级学生的 18 年（2004—2021）同名课程教学实践编写而成，适合作为高等农林院校同名课程的双语教学用书，也可用于农业院校和科研院所研究生考试的参考书，以及作为全国农业生态化转型背景下的生态农业管理和实践者的参考书。

　　我们对在本教材编写和出版过程中付出辛勤劳动的作者和编辑致以真诚的感谢，特别感谢国际版编辑、德国施普林格（Springer）出版集团 Raagai Priya Chandrasekaran 女士和国内版编辑、科学出版社丛楠女士的精心帮助。同时也感谢资助两位主编从事合作教学的中国和加拿大相关政府部门和组织。

　　本教材获得福建省社会科学规划项目（FBJ2018B070）、福建省创新战略研究项目（2020R0138）和福建农林大学教材出版基金项目（111971702）的资助。

王松良

2021 年 3 月 9 日于福建农林大学金山校区

Contents

Part I
Context of Agroecology

Chapter 1
Agriculture and Its Anthropocentric Sciences

S. Wang and C. D. Caldwell

"…but just simply distributing seeds and fertilizer, if that's the plan, it's going to fail long term."
—Howard Buffett

Abstract The current model for industrialized agriculture is at a crossroads, characterized by degraded soil and environment, unbalanced human nutrition, food safety issues, loss of trust between consumer and producer, and finally sustainability at risk. By critically reviewing and integrating the definition and technological evolution of agriculture, agroecology links agricultural sciences with ecology to form an interdisciplinary and holistic science, generalized for steering an imperative agroecological transition in this anthropocentric era.

Learning Objectives

After studying this topic, students should be able to:

1. Explain the concept of the Anthropocene Era and why the earth may now be considered a "human planet".
2. Describe three current important topics affected by how we do agriculture production.
3. Define the terms agriculture, agricultural sustainability, and agroecology.
4. Discuss how the discipline of agroecology has evolved and fits in the philosophy of agricultural sustainability.
5. Describe the three pillars of agricultural sustainability.
6. Explain why corn is a good symbol for the study of agroecology.

1.1 Introduction and the Anthropocene Era

In May 1987, Jared Diamond published an article in Discover Magazine, entitled "The Worst Mistake in the History of the Human Race" in which he puts forward the thesis that agriculture, despite all its apparent positive effects on health, well-being, and survival of human populations, may have been a very negative turning point for humankind. He sees that agriculture has in many ways produced the disaster of social problems, disease and dysfunctional society. Nowadays, we would also add to that list of negatives associated with agriculture, problems of environmental degradation. Whether we agree with Dr. Diamond's conclusions or not, human needs have certainly reshaped global ecosystems.

Where previously there was a hunter–gatherer society, a drastic change occurred when farming was invented about 10,000 years ago. That is only about 300 generations ago and the world population was 1 million. About 5500 years ago, we developed cities and civilizations and the world population was about 5 million. Along came the Industrial Revolution about 150 years ago and the broader population had reached 1 billion. With World War II and the great acceleration 70 years ago, our population has blossomed. It took 50,000 years to reach the first 1 billion people; the last billion took 10 years.

Technically, geologists would say that we are in the geologic Era called the Holocene. This is the period of the past 10,000 years of reasonably stable climate since the last Ice Age. We learn a lot from looking at layers of rock to determine the past geology of our planet. By examining the fossil record we can see the coming and going of species throughout the ages and speculate on what made some species survive and others not. It is interesting to speculate on what people in the future will tell about our time on this earth. From what we know at the moment, the geology will tell a story of extinct species, changes in the ocean chemistry, fewer forests and more deserts, dammed rivers and retreating glaciers, sunken islands, aluminum drink cans and plastic bottles, and perhaps large evidence of megaprojects like the oilsands in northern Canada. There are some geologists who, considering the present state of the world, would call this new era the Anthropocene; this is in recognition that humans have become geophysical forces with effects as much as asteroids and volcanoes have had in the past.

If we consider that earth is now a "human planet", we need to consider how agriculture has contributed to this new era. The invention or evolution of agriculture 10,000 years ago was perhaps triggered by climate change at the end of the Ice Age. Warmer, drier conditions were conducive to the growing of cereals (annual grasses) which were extremely successful due to their ease of production, their concentrated nutrition, ease of storage and transportation. This new culture depended on burning, shifting cultivation. However, with the domestication of chickens, pigs, and cows, there was an opportunity for the cycling of nutrients by putting manures back on the ground. This natural fertilizer worked well within such systems. However, as populations concentrated in groups, these natural fertilizers were not enough to avoid the starvation of masses. This led to the invention and widespread use of artificial fertilizers.

The next steps in agriculture included extensive land clearing, increased use of water for irrigation (note that we now use 70% of our freshwater for agriculture), and a large increase in both cereal production and domesticated animals. Recently, as there has been a growing middle class throughout the world, there is an increasing demand for meat and milk. As billions of people want more of these precious commodities, more of the earth's resources are strained. It is an interesting statistic that the mass of cows on earth is now 1.5 times the mass of all humans combined.

In 1974, the first world conference on food was held. The promise at the end of that conference was that in 10 years' time no child would go to bed hungry. This was an ambitious goal that we still have not met. The Green Revolution in the 1960s and 1970s was a great success in saving millions from starvation. Its success was on the back of better plant breeding, the use of targeted pesticides, increased water usage, and inorganic fertilizers. It boosted food security for millions of people at the time. The Green Revolution also brought great disruption to social and economic stability and has had long-term impacts on the environment. However, now we still have nearly 1 billion people that do not have enough to eat (and 1.5 billion that are fat).

The challenge in the Anthropocene is that we need to double food production by 2050. This seems to be an overwhelming challenge; however, we are the solution. We need to recognize both our mistakes and our successes in the past and learn from them. The hope of this textbook is to help educate ourselves, then think imaginatively and work together. Every person and every action count.

Obviously, we cannot turn back the clock, all become hunters and gatherers, scattered societies living peacefully unaware of each other. However, the lessons that Dr. Diamond has pointed out live on today. It is the role of agroecologists and of the educated society to use the principles of good agriculture to have positive effects on human health, well-being, society, and our environment. There is a need to build on the positive aspects of agriculture and what we have learned over the last hundred years and apply it to new systems, new systems that will become sustainable models for human coexistence with the living world around us.

1.2   What Are the "Hot" Issues in Agriculture and Food System?

Agriculture, the human activity from which almost all our food originates, has been a major concern for humanity throughout recent human history. Despite the great achievements that have been made in food production since the Green Revolution period of the 1960s and 1970s, the challenges both in quantity (security) and quality (safety) of food remain imperative.

By 2050, the increase of an additional 3 billion people, coupled with the increasing consumption per person, means agricultural production will need to expand by 60% to meet increasing demand (Lipper et al., 2014). It is an urgent challenge because the infrastructure of agriculture has been heavily degraded by inappropriate farming practices to reach our present food production goal. During the first Green Revolution of the 1960s and 1970s, the excessive use of synthetic fertilizer, chemical pesticides, and other fossil fuel-dependent chemicals has caused degradation to the soil and water. This is not just a scientific problem but rather a problem of global society, poverty, and human equity. Seventy-five percent of the world's poorest people are living in ecologically vulnerable areas, where agriculture is not only the food source but also crucial to poverty reduction.

At present, conventional agriculture is responsible for detrimental changes in land use and greenhouse gas emissions that lead to climate change. Among 4.0 billion hectares of earth's surface land, 38% was used by agriculture, producing 30% of global net production for the human diet (Zabel et al., 2019). However, this has been achieved in many cases via inappropriate farming practices; e.g., monoculture and an industrial agricultural model which is been a major force for global biodiversity loss (Zimmerer et al., 2019), nearly 80% freshwater use (Liu and Song, 2019) and 20%–30% of greenhouse gases emissions (Uphoff and Thakur Amod, 2019). Hence, agriculture, which is a unique life industry, confronts an unprecedented challenge. We need to convert the prevailing model of industrialized agriculture into a new vision of agroecological agriculture:

1. Working with the soil to increase its overall health thus recovering its potential productivity. The use of synthetic fertilizer has contributed much to crop yield, but the additive effects of more than 50 years of excessive use of synthetic fertilizer has led to soil degradation and nutritive deficiency in our plant food system. It is imperative to reverse this trend and return to a status of soil health to secure the food system.

2. Diminishing the climate change impact of farming to produce a resilient, adaptive model. A well-balanced agricultural system can be both carbon emitter and sink, but the prevailing industrialized agriculture, with its dependence on fossil fuels for the production of fertilizers and pesticides, tends to deplete the carbon sink in the soil leading to the prevailing carbon emission status. It is time to rediscover agriculture's carbon sink characteristics.

3. Focusing on the positive goals of farming to maintain human health and recover biodiversity in the world. Under the older reductionist industrial model of agriculture, there was an idea that we needed to eradicate pests completely from the systems to increase yield. We now know this to be both inappropriate ecologically and to be counterproductive both environmentally and economically. This overuse of chemicals has resulted in not just eradicating some pests but has led to decreased biodiversity. The loss of biodiversity, degradation of the environment, and decrease in food safety needs to be reversed by a change in ideology.

4. Moving from a reductionist ideology to a philosophy of sustainability. The specialization strategy of modern industrialized agriculture has focused on single commodities that encourages "quick fix" technical solutions from agricultural sciences for the short term, undermining in many cases the sustainability in the long term. Hence, it is urgent to use agroecology as a holistic and interdisciplinary framework for agricultural research and education to steer an agroecological transition worldwide.

1.3 Agriculture, Agricultural Sustainability, and Agroecology

In general, agriculture is a misunderstood science and practice, tainted with the ideas that the endeavors of those involved are solely self-serving. In fact, the Oxford Concise Dictionary dismisses agriculture as nothing more than, the "science or practice of cultivating the soil and rearing animals" (Oxford Dictionary Editorial Board, 2019). This cursory definition only explores one facet of the complex study and practice of agriculture.

1.3.1   Evolution of Definitions of Agriculture

As one considers the various activities and impacts of agriculture, our functional definition must evolve. Outdated ideas of agriculture depict this key human endeavor as only one of raising animals and plants for human use. The various skills and knowledge required of someone in agriculture traditionally included an understanding of soils, how to grow and process crops, and animal husbandry including both ruminants and non-ruminants and agroforestry.

Over time, our understanding and concepts of agriculture and its role in civilization has evolved. Agriculture played a key role in population settlement and the rise of modern civilization. The husbandry of domesticated plants (i.e., crops) and animals allowed for these societal changes, creating food surpluses that enabled more densely populated and stratified communities.

A general agronomic textbook in China defines agriculture as "the infrastructure of national economy, and main source of living, as well as industrial material for human beings" (Zhai, 1999). This definition regards agriculture as a source of food, clothing, and financial income, focusing on agriculture as "money-based" and economy-driven. Agricultural practices under these defining ideologies have deteriorated natural resources and the environment (soil, water, and atmosphere). This has become apparent by the depletion of ecological diversity and productivity from increased agricultural production, to meet an ever-increasing demand for food by an overpopulated society. The destroyed ecological diversity and productivity can no longer support infrastructure and provide a base for all agro-economic productivity. The Industrial Revolution replaced many organic materials with synthetic materials in agricultural ecosystems (agroecosystems), which led to the formation of "Modern Fossil Fuel Agriculture". This increased dependence on synthetic materials and fossil fuels in agriculture has only exacerbated the deterioration of agroecosystems. However, agriculture is multifaceted and extremely important; it is not limited to food production and economic output, but rather, it is a critical interface between humans and nature.

Ancient Chinese books, poems, and paintings depicted an ideal farming-oriented society among a charming agroecological environment. The ecological functions of sound agriculture and forestry, as depicted in ancient Chinese art, are summarized as follows (Wang, 2005):

1. Water and soil conservation.
2. Climate and rainfall regulation.
3. Protection of land from wind and desertification.
4. Landscape beautification and pollution prevention.
5. Living energy supply and source of fertility.

Fortunately, researchers and government agencies in both developed and developing countries are gradually recognizing the multi-functionality of agriculture.

Therefore, Caldwell (1996) redefined agriculture as "the science, art, politics and sociology of changing sunlight into healthy, happy people". This expanded definition better recognizes the function of agriculture; it means that agriculture is a natural ecological process, whereby solar radiation converts energy and matter from natural resources (including land, atmosphere, and water) for food and human environment. This definition incorporates people as an essential component, emphasizing the equilibrium point for human benefits and natural conservation, or "the basic interface between people and their environment". One can group the major agricultural products in the following categories: foods, fibers, fuels, raw materials, pharmaceuticals and stimulants, and an assortment of ornamental or exotic products. Ecological designs of agronomic and horticultural systems have become part of the functionality of agriculture.

The point to note here is that agriculture is homocentric. A literal interpretation of which would put people at the center of agriculture; a functional interpretation puts people as a beneficiary but not central. Such a holistic notion fits in the meaning of the sustainability of agriculture, and thus the sustainable food system.

1.3.2   *The Science of Agriculture and Agricultural Sustainability*

Agricultural sustainability or sustainable agriculture originated from Sustainable Development (SD), which was defined by the World Commission of Environment and Development (WCED) of UN in 1987 in its famous report "Our Common Future", as "the type of development allowing the current generation to satisfy its needs without the risk of depriving future generations of this possibility" (WCED, 1987). The term SD emphasized carrying on from generation to generation, in other words, "sustainable" means "intergenerational"; as such, agriculture is the original human industrial sector that must be sustainable (SD is impossible without food security), and also could be sustainable for its natural origin (linking food security to the science of Ecology).

Sustainable agriculture was framed by the World Bank's Consultative Group on International Agricultural Research (CGIAR) in 1988 (FAO, 1988). According to Jules Pretty, a sustainable agriculture should have the following properties (Pretty, 2008):

- Farming practices that do not harm the environment.
- Be practical and cost-effective for farmers.
- Improve food productivity.
- Have a positive impact in terms of environmental health and services.

In sum, sustainable agriculture should be resilient in response to off-farm stresses and shocks and be able to persist and sustain productivity and social/environmental stability over long periods.

In addition to supporting the industry and the economy, agriculture also promotes land renovation, biodiversity, nature conservation, and design.

1. Agriculture is an economic engine globally.

Supports the livelihoods and subsistence of people worldwide.

2. The agricultural sector must simultaneously:

(a) Produce sufficient excellent quality goods to meet demand.

(b) Be a positive force for the maintenance of biodiversity and the natural environment.

(c) Enhance human health of diverse populations across a widening economic spectrum and social background globally.

Hence, agricultural management must continually increase the productivity of existing farmland to meet population demand through the adaptation of good and efficient management practices. Additionally, management should embrace the three pillars of agricultural sustainability, representing natural (environmental), social, and economic aspects (Fig. 1.1).

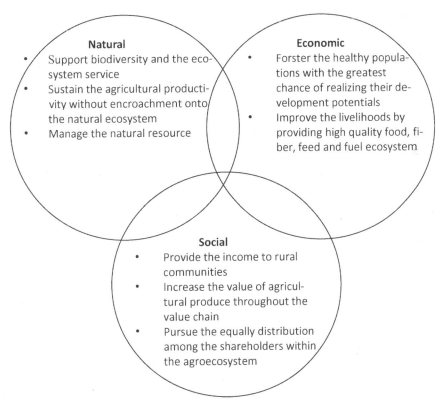

Fig. 1.1　The three pillars of agricultural sustainability

Science and technology are the building blocks of modern agriculture. One must understand the biological and physical sciences underlying agricultural engineering and technology. Successful farming requires knowledge of tillage, irrigation, fertilization, drainage, and sanitation. Some aspects of farming require further specialized knowledge, which engineers and other specialists apply in agricultural systems. Agriculture encompasses a wide variety of specialties and techniques.

One such specialty is the ability to increase suitable land for plant production, usually performed by digging water-channels and other forms of irrigation. Cultivating crops on arable land and pastoral herding of livestock on rangeland are some of the fundamental practices of agriculture.

In the past few decades, plant breeding, agricultural chemistry (e.g., pesticides and fertilizers), and corresponding technological improvements have sharply increased yields from cultivation (Table 1.1). For instance, plant genetics and breeding contribute immeasurably to farm productivity; meanwhile, genetics have turned livestock breeding into a science. However, some of these technologies may cause widespread ecological damage and negatively impact human health (Pretty et al., 2001). Hydroponics, a method of soilless gardening in which plants are grown in chemical nutrient solutions, can help meet the need for greater food production as the world's population increases. Similarly, selective breeding and modern practices in animal husbandry, such as intensive pig farming (and similar practices applied to chickens), have increased the output of meat. However, hydroponics can lead to pathogen attacks and selective breeding in crop varieties has led to the utilization of only a few plant species and monocropping, reducing biological diversity. In addition, concerns have increased about animal welfare and human health effects from antibiotics, growth hormones, and other chemicals often used in large-scale meat production processes.

Agricultural chemistry which includes, but is not limited to, the application of fertilizer, insecticides and fungicides, soil makeup, analysis of agricultural products, and nutritional needs of farm animals, must consider many crucial farming concerns. The increasing use of inorganic fertilizers and synthetic pesticides poses many problems in soil degradation, groundwater contamination, food safety, toxicity accumulation in natural wildlife, and other environmental deterioration.

Table 1.1　Global productivity comparison during three phases of agricultural development (Wu, 1986)

Phase of agricultural development	Land productivity $(J \cdot cm^{-2})$	Number of people fed per hectare of arable land	Rural population (%)	Purchased rate of agricultural product (%)
Historical agriculture	0.054×10^6	20	100	0
Labor intensive agriculture	1.020×10^6	280	85.7	14.3
Modern agriculture	4.180×10^6	1000	6.4	93.6

The packing, processing, and marketing of agricultural products have also been influenced by the sciences and technologies. Methods of preservation, quickfreezing, and dehydration have increased markets for farm products and decreased post-harvest losses. These processes do, however, mean the use of more chemicals and materials potentially leading to resource depletion, food unsafety, and increased environmental pollution. However, it is absolutely crucial that postharvest losses are decreased since this is a critical loss of nutrition and income in the food system.

Historically, instruction in the agricultural sciences has emphasized the understanding and enhancement of production, maximizing net returns on single products per unit of land or labor.

This is often been to the detriment of the environment.

Many problems resulting from modern agriculture occur because of reductionist disciplines and utilitarian technologies. Thus, we need to modify our understanding of agriculture, integrating Caldwell's newer definition (1996) with a new science and discipline to investigate agriculture in a more inclusive way (Wang, 2005). Agroecology, therefore, should be considered as a modern discipline that bridges ecology (including human ecology) with agriculture.

1.3.3   The Link to the Discipline of Ecology

The etymology of ecology stems from the Greek words "oikos" (house or place to live in) and "logia" (study of). The word ecology was proposed and defined by the German biologist, Ernst Haeckel, in 1866. His definition stated "Ecology is the science of the relations between organisms and environment" (Odum and Barrett, 1977). This definition implies that ecology builds upon related biological sciences such as zoology and botany; such disciplines usually examine organisms themselves, whereas ecology explores the relationships among organisms and their environment. While ecology can be considered a biological science, it spans a much broader study area, including earth science, chemistry, physics, mathematics, medicine, and certain aspects of the social and economic sciences. The famous ecologist, Eugene Odum, stated that Ecology is "a science bridging biology and social science" (Odum, 1975). This explains ecology in terms of an interdisciplinary science, mixing natural science with social science, where one can also infer an emphasis on economics and politics. A holistic or integrated approach to the investigation of ecosystems requires considerable knowledge, effort, and scientific resources. The results of ecological studies are often contrary to what one may expect from a reductionist approach.

Although the study of ecology traces back to ancient Greek and Roman times, modern ecology was born from and accelerated by the social and environmental problems of the eighteenth-century Industrial Revolution. Modern ecology originated as a response to the global emergence of the "Five-Ecological-Crises", i.e., Population, Food, Resource, Energy, and Environment, at the beginning of the twentieth century (Meadows, 1972). Essentially, ecology is the economics of nature, as opposed to the money-based economics that overshadows the social economy. Economics focuses on accounting for ways to regroup resources to maximize the output, regardless of any abstract innate value.

Ecology investigates the interactions among organisms and their environments at various scales, from individual organisms to population, community, ecosystem, and biosphere.

1.3.4   The Link Between Agriculture and Ecology: The Inclusive Discipline of Agroecology

The role of ecology in agriculture is to find the pivotal balance among global food security, advantageous production, technological innovation, environmental preservation, protection of biodiversity, and economy. Both agriculture and ecology have common roots in the disciplines of botany, chemistry, physics, and geology, with very distinct applications and management practices. Agroecology evolved from these relationships, first emerging in the 1930s; its initial phase lasted until the 1960s, after which the study expanded until it became considered a discipline in its own right and institutionalized in the 1990s (Wezel et al., 2009). Luo Shiming et al. (1987) defined agroecology as "a science of the interaction, co-evolution, regulation, control and equilibrium development between agro-organisms and their environment (both natural and social), based upon the principles of ecology and system theory and practice". This definition is inclusive and reflects the intent of Caldwell's definition (1996) of agriculture. While agroecology is derived from the larger field of ecology, agroecology draws even more strongly on the social sciences to construct understanding and predictions about organism and environment relationships. Ecology can be considered the "parent" theory of agroecology because the goals of the discipline are to pursue the sustainable management of particular ecosystems, i.e., agroecosystems. This has been evidenced by the long history of corn (maize, *Zea mays*) in evolution in the agroecosystem (see Sect. 1.4).

1.3.5　The Development of Agroecology as a Discipline and Science

1. Agroecology developed under the background of the global ecological crises concerning agriculture.

　　Agroecosystems play a crucial role in our lives because they provide us with food and fiber while greatly impacting the quality of our environment. Historically, global ecological crises concerning agriculture have been the inspiration for the development of agroecology, as well as a major source of conflict. China represents one of the ancient farming societies of the world. In China, nature has been destroyed mercilessly since the first hoe was used in agriculture, transforming the "Green Grassland" into the "White Desert", and "Mother River" into the "Yellow River"(Wang et al., 2019). Globally, ecology has been involved in revolutions both of environmental protection and of environmental sciences. Soon after Rachel Carson (1962) revealed the far-reaching effects of chemical pesticides used broadly in agricultural practice in her famous book "Silent Spring", many ecologists dedicated their careers to agroecosystem research.

2. Agroecology promotes the evolution of the agricultural economy from "industrial-based" to "livelihood-based" economy.

　　"Historical agriculture" refers to the first 10,000 years of agricultural development. Prior to that, people lived by hunting and gathering. Historical agriculture required extensive cultivation and productivity relied entirely on the unpredictability of nature and weather. This economic pattern could be realistically defined as a "nature-based economy". Subsequently, labor-intensive agriculture was born 2000 years ago, the productivity of which depended totally on people power and land characteristics. This economic pattern could be described as a "subsistence-based economy". The agricultural pattern has changed drastically from a "nature-based economy" to an "industrial-based economy" since the Industrial Revolution. Modern agriculture can be formularized by "fossil energy + technology = commodity", showing that fossil energy and modern technology have become the dominant factors for agriculture. Consequently, agricultural productivity has increased tremendously when compared with the former two stages (see Table 1.1). As a negative result, however, such patterns have generally destroyed the natural sustainable basis of agriculture.

By the 1980s, developed countries started responding to the destructive aspects of modern agricultural patterns with several alternative agriculture such as organic, ecological, biological, and natural agriculture. Since the 1990s, the concept of Sustainable Development has arisen as common sense around the world, allowing for the emergence of Sustainable Agriculture. Sustainable Agriculture is characterized by dependence on "intelligence" (knowledge, technology, and education), placing people's livelihood over monetary return; it could therefore be called an "intelligence-based economy" or "livelihood-based economy". The primary purpose of Sustainable Agriculture is to realign the relationship between agro-organisms and the environment. As a scientific approach to sustainable agriculture, agroecology aims to study, diagnose, and offer alternative low-input management strategies of agroecosystems (Altieri, 1995).

3. Agroecology: a new research and development paradigm for world agriculture.

In the last 100 years, world agriculture has been driven by the application of science and technology to achieve the tremendous increases in productivity to respond to the increased needs for an adequate food energy supply for the increases in world population. Improved plant and animal breeding has resulted in more productive cultivars of crops and breeds of animals with more efficient use of inputs. At the same time, there has been an increased dependence on chemical use. This includes synthetic fertilizers, pesticides, and in some cases plant and animal hormones. Many countries benefited immensely from such industrial agriculture; however, along with these increases in productivity have come negative impacts on the environment. There has been environmental price to pay for monocropping, excessive tillage, short rotations, and overuse of chemicals in severe soil degradation. Starting in the 1960s, a huge international effort began, the so-called "Green Revolution", which extended a few highyielding crop varieties to many countries in order to achieve adequate world food calorie/protein production. However, this process depended on water, fertilizer, and monocropping, resulting in biodiversity loss and socioeconomic upheaval.

The global environmental revolution was initiated by the response to "Silent Spring" in 1962. Rachel Carson exposed the indiscriminate use of toxic chemicals in the "modern" agricultural industry. More and more people expressed concern about the "modern" agricultural paradigm that threatens the natural resources of land, soil, air, and water. These resources are threatened through processes such as loss of soil fertility by erosion, acidification, salinization, and desertification. A new model was needed, and eventually, the sustainability paradigm was advocated worldwide.

As an introductory discipline of sustainable agriculture, agroecology was proposed as a new research and development paradigm for alternative, sustainable agriculture (Altieri, 1989). Agroecology advocates using ecological principles when growing crops and raising livestock; this involves decreasing the ecological footprint throughout the full value chain. At the production level, this includes the increased use of such techniques as organic fertilizer, biological pest control, soil conservation, and in general, limiting the use of nonrenewable fossil fuel energy.

Agroecology was proposed as a new scientific discipline that defined, classified, and studied agricultural systems (Gliessman, 2005). Agroecology was developed to give scientific and socioeconomic guidance to the management of agroecosystems (Conway, 1985). Therefore, going beyond the reductionist approach, agroecology provides a methodology to diagnose the "health" of agricultural systems and thus guide the design of sustainable production systems in a more fully integrated manner.

To manage an agroecosystem sustainably, agroecology uses a goal-oriented design using various methods and appropriate input technologies to achieve those goals. This approach is designed to improve soil health, decrease waste, enhance biodiversity, diversify farming practices, and respond to socioeconomic demands. It is therefore crucial for a team of agricultural scientists and social scientists alike to be involved in the search for sustainable agricultural technologies.

1.4   Corn as a Symbol of Agroecology

Throughout the cultural, political, economic, social, and gastronomic history and development of people, no species has more closely aligned with us than corn (*Zea mays*). This is explored more fully by Ruben G Mendoza (2003) entitled "The Natural History of Maize".

Several important points discussed in Mendoza's paper link maize and people together:

1. Maize is dependent on people for its survival as a species and, it can be argued, that people are dependent on maize for survival. While maize is a very successful crop, it is not a successful wild plant. Without the assistance of people, maize would become a very marginal plant and perhaps even disappear from the earth. Our breeding and selection efforts over millennia have produced a super-crop, but as an individual plant has poor species survivability. One needs to look no further than the seed covered cobs in one location to see the problems in independence for the plant. On the other hand, the worldwide demise of maize would decimate the human population due to widespread famine and resultant social disorder. While people would not disappear, our numbers would drop and our lifestyle would change considerably.

2. Agroecology deals with how people change their environment to benefit themselves. Maize is an excellent example of a crop that has co-evolved with people over the millennia. Maize has

changed from a wild, relatively unproductive crop to one that can adapt and respond to many forms of human management, from low-tech to high-tech. The first maize selections made by primitive peoples set the stage for generation after generation of "improvements" to the crop, followed by targeted breeding and crop management in the development of hybrids. This means that people in various places across the globe have improved the ecotypes to respond to both climate and management so that the crop of maize is widely adapted as a species but narrowly adapted as a hybrid.

3. Agroecology addresses the range of interactions of humans, plants, and animals; maize is a key component of both human food and animal feed.

4. Agroecology deals with closing the nutrient cycle and maintaining balance within systems. As a symbol of agroecology, maize is responsive to nitrogen of any source but is particularly good at utilizing animal manures.

5. Maize is given both praise and scorn. Maize is praised as a major source of food for people and feed for animals, while it is negatively targeted by the "Green Revolutionaries" as anything but "Green". Maize was a key component in the global change in agriculture, bringing with it new seeds, inorganic fertilizers, and pesticides to areas of the world that had not had them before. These changes produced higher yields but also induced social and environmental disruption. Agriculture and maize have worn the mantle of scorn for that. Agroecology seeks to rehabilitate the image of agriculture and maize will continue to be a key component of that agriculture as it coevolves with us.

Chapter 2
Natural Ecosystems Versus Agroecosystems

C. D. Caldwell

"If you do not supply nourishment equal to the nourishment which is gone, life will fail in vigour, and if you take away this nourishment, the life is entirely destroyed."
—*Leonardo Da Vinci*

Abstract This chapter uses a comparison of natural ecosystems with managed ecosystems to explore the concepts of energy and matter flow. Basic concepts of ecological efficiency help inform how we can best produce food for our population using the least land needed with the lowest environmental impact. An initial exploration of the concept of ecological footprint will assist students in the process of critically reading chapters on our present agricultural production practices.

Learning Objectives

After studying this topic, students should be able to:

1. Explain the concept of ecosystem and how it applies to agricultural ecosystems (agroecosystems).
2. Compare Natural and Managed (people-centred) Ecosystems, using examples, in terms of:

   (a) Open versus closed systems
   (b) Energy amounts and flows
   (c) Nutrient amounts and flows
   (d) Diversity and stability

3. Explain the concept in agricultural ecosystems of "Harvest the Sun".
4. Explain the relationship of "Harvest the Sun" to LAI, LAD and crop yield.
5. Explain the concept of ecological footprint.

2.1 Expanding Our Understanding of "Ecology"

As was discussed in Chap. 1, "ecology" is the study of the relationships between organisms and their environment. Therefore, ecologists, those people who study those relationships, define a study area as an "ecosystem". These are the particular biological communities and the physical environment in which the interactions occur. The largest defined ecosystem is the biosphere, which includes all living organisms on the Earth and their interactions with our environment. This is the global ecosystem, with all its complexities, that is the unit of study for many of our climate change scientists. However, most ecosystems are defined in a much smaller space and time. For example, one could define an abandoned old field as an ecosystem with its complex of many types of plants, invertebrates, vertebrates, bacteria and fungi all interacting with their abiotic environment. Similarly, many students have used a small pond in the original studies of ecosystems and been surprised by the plentiful life that can be found there.

Therefore, most students are very aware of this definition of an ecosystem, including the idea of a community of all of the populations of all of the species as the biotic, or living part of the ecosystem. This is complemented by the abiotic component; i.e. the nonliving or physical environment. We define the boundaries of an ecosystem in order to better understand the interactions. The concept of an ecosystem is applied to agricultural systems to define a unit of study which we would call an agroecosystem. A convenient and practical definition of an agroecosystem maybe a farm in its entirety. This allows one to apply the principles of ecology to better understand the interactions and possible synergies that can be drawn from such an analysis. A smaller unit might be a particular production field that would allow for specific analysis of the interactions and impacts of management of that unit. A larger agroecosystem might be defined as a watershed; several farms within a watershed will have complex interactions and impacts on each other and the environment. Other options for a definition of a particular agroecosystem are possible; the definition of the agroecosystem depends on the purpose for which the study is intended.

The concept of an agroecosystem comes directly from our understanding of what is agriculture. A typical definition of agriculture would be the science or practice of cultivating the soil and rearing animals. However, this does not capture the true nature of agriculture. To understand that we need to understand the interface between people, food and the environment. We expect agriculture to provide many things to us: safe food, enough food, a diversity of food and food products, affordable food, an attractive countryside, income generation and wealth. There are also many things that we expect agriculture *not* to do: poison the water, pollute the soil, contaminate the air, cause social problems or unrest, or be an economic burden on the state.

Considering all of the above, it is obvious that agriculture is a science; i.e. it is a systematic enterprise that creates, builds and organizes knowledge in the form of testable explanations and predictions. It is also obvious that agriculture is an art; i.e. it is something that is created with imagination and skill and that is beautiful and expresses important ideas or feelings. Anyone who deals with agriculture either on a local or larger level also knows that agriculture is tightly connected to politics; i.e. agriculture deals with the structure, organization and administration of the state (Aristotle). In addition, agriculture is very much at the interface of rural and urban societies and the understanding of agriculture requires an understanding of social structure. Therefore, agriculture is also part of sociology; i.e. the systematic study of the development, structure, interaction, and collective behaviour of organized groups of human beings. Perhaps the final part of being able to define agriculture is to consider the goal. The goal of agriculture is indeed to produce healthy, happy people.

This leads us then to this logical definition of agriculture: *agriculture is the science, art, politics and sociology of changing sunlight into healthy, happy people.*

With this broad understanding and concepts of agriculture, one can look at what an agroecosystem truly is. Early on in our education, we have been taught that an ecosystem is a living community that includes all the factors in the nonliving environment along with the interactions among both biotic and abiotic factors. Thus, an agroecosystem is that broad-based type of ecosystem which is homocentric in nature and branches across disciplines in our understanding. Agroecology must therefore deal with the observations and analysis of our human-centric agricultural systems as they interact both with the biotic and abiotic environment.

2.2   Natural Versus Agricultural Ecosystems

Fig. 2.1 is a picture of a natural setting with minimal direct human contact. For discussion purposes, let us consider this to be a "natural ecosystem". The picture shows the diversity of plant life including a canopy of trees, shrubs and varied understory of greenery. If one were to walk through this forested glen, it would be obvious that there is an abundance of life. You would perhaps see small vertebrates, such as mice and moles, maybe some rabbits and their predators like the fox if you walked carefully through it. Looking closely in the understory, one can see a multitude of insects in various stages of

Fig. 2.1   A picture of a natural setting with minimal direct human contact (photo credit: pixabay.com)

development, deposition of small eggs, various larvae, maybe hundreds if not thousands of different types of crawling burrowing or flying insects. Digging into the earth, again we find a plethora of life; earthworms, grubs, mycorrhizae, various other fungi, bacteria and plant roots intertwined with each other, exchanging material and information. If one looks carefully at such a natural ecosystem, the life and its interactions are overwhelming.

The first characteristic of this ecosystem, therefore, is diversity; it has evolved with multiple food webs and failsafe interactions. If one part of the web is damaged or severed by disease or predators, another part of the food web takes over. Diversity means stability and resilience to change. The second thing one may notice in this natural ecosystem is that it is a relatively "closed" ecosystem; i.e. lifecycles of those organisms in the ecosystem are completed in that physical place.

Organisms are born, live, die, decompose and provide nutrients for the next cycle of life all in the same place. Obviously, no ecosystem is completely closed; if a hawk comes through and grabs one of the mice on the ground and takes it off to another place, this is a movement of matter and energy out of the system, meaning it is somewhat "open". However, for the most part, one can say that natural ecosystems are mostly "closed".

Considering the plants that are present in this ecosystem, as mentioned above, they are of the wild type. This means that their photosynthetic rate per unit area is lower than what would be expected from domestic plants; the overall total energy capture rate is relatively low. However, the energy flows essentially within the closed system and cycles with the lifecycles of those organisms within it. The nutrient amounts found in the soil are perhaps lower than one would expect if you are used to a farm soil; the total amount of nutrients in the system may be very high but, for the most part, the nutrients are held in the biomass. It is only when biomass becomes dead material that the nutrients join the cycle. This gives a large distribution of nutrients throughout the ecosystem.

In summary, this natural ecosystem is mostly closed, has relatively low energy amounts captured per unit area per time, and the nutrients are mostly held in the biomass. However, one of the great strengths of this ecosystem is that it is highly diverse, highly stable and resilient to change.

Fig. 2.2 helps us consider a managed ecosystem or agroecosystem. One can easily see that the system has been simplified compared to that of a natural ecosystem. Rather than having a multitude of different small herbivores, the ecosystem is dominated by one large type of herbivore, the cow. Looking at the vegetation, it is quite simplified, being reduced to just a few species, the tastiest, nutrient-rich and most productive ones for those herbivores. Therefore, the first thing to note is that this is a lower diversity ecosystem and therefore has the problem of lower stability and resilience in the face of disturbance. In favour of this ecosystem, however, is the very high productivity. The domesticated plants in this picture are highly efficient photosynthetically and are managed in such a way that they capture huge amounts of energy per unit area per time. Consider now whether this is an open or closed system. The cattle at the end of their time on this pasture will either produce milk to be sold off the farm or meat to be taken off the farm. Either way, there is a huge amount of matter and energy leaving the ecosystem. This is a very open ecosystem; energy and matter come into it in the form of fertilizers, diesel fuel, human labour and so on and leave the farm in the form of products. The energy amounts are high and the flow is in and out; nutrient amounts are high and the flow is in and out. If one looks at where the nutrient stores are, to a large extent they are in the soil, as a result of fertility treatments to favour the fast growth of the vegetation, needed to promote the large herbivores in the system.

Fig. 2.2   Helps us consider a managed ecosystem (photo credit: pixabay.com)

Therefore, the agroecosystem is an open ecosystem, with high energy amounts and flows, high nutrient amounts and flows, and low diversity/stability. Human manipulation and alteration of land for the purpose of establishing agricultural production differentiates agroecosystems from natural, unmanaged ecosystems. Therefore, although energy flow initially generates from solar radiation in both systems, human-driven agroecosystems have much greater energy subsidies from the socio-economic systems.

For self-sustainability, agroecosystems should optimize solar energy and fuel internal trophic interactions to maintain system function. In many developed countries, the evolution of energy subsidies in food production systems has greatly changed since the beginning of the twentieth century when industrial agriculture started to prevail.

Farmers are the original ecologists, and most have a sophisticated appreciation of these issues. Many farmers work to increase the diversity of crops by using such treatments as intercropping, forage mixtures and complex rotations. However, these do not approach the diversity of a wild natural ecosystem. Similarly, many farmers introduce nitrogen-fixing legumes which take nitrogen gas from the air and change it into nitrogen fertilizer and manage animal wastes in an efficient way; this decreases the off-farm input of inorganic fertilizers. The recognition by farmers of the limitations of the managed ecosystem stimulate innovative farmers to do better. In particular, there is a recognition that the soil itself is a living ecosystem. The presence of large, healthy, active organisms like earthworms is a good indicator of soil health and increasingly farmers are using soil health as an indicator of positive management, not just the indicator of profit. There is a reason for optimism. There is a willingness among farmers to find better methods, transfer new appropriate technology with lower environmental impact and to practice the art and science of growing food with goals that are long term and beyond profit.

2.3　Energy Flow and Matter Cycle in a Natural Ecosystem

Solar energy flows into and through ecosystems powering the nutrient cycles that maintain system stability. Solar energy initiates photosynthesis by plants (producers), converting solar energy from its radiant state to stored chemical energy (carbon-based matter). Several trophic levels of consumers use the stored chemical energy in the food chain and food web; decomposers recycle the remaining waste material. Decomposing organic chemicals release carbon dioxide back into the atmosphere and are converted into inorganic chemicals that producers in the next part of the life cycle absorb. The amount of energy converted depends on the efficiency of plants, consumers and their abiotic factors. Producers can use the recycled matter as the inorganic building blocks of the living system repeatedly if they are not washed away or removed from the ecosystem. Fig.2.3, 2.4 and 2.5 demonstrate in different ways how food chains and food webs interact in terms of energy flow and ecological efficiency.

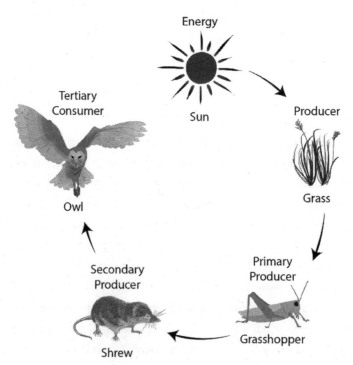

Fig. 2.3　Example of a simple food chain (photo credit: S. Mantle)

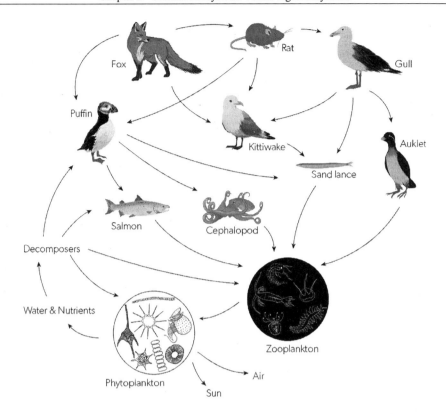

Fig. 2.4   Example of a more complex food web (photo credit: S. Mantle)

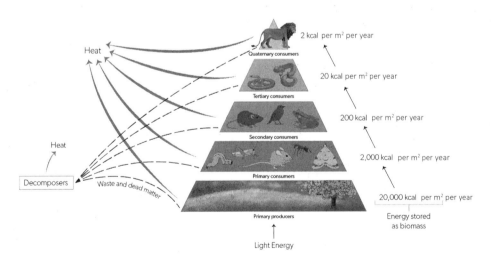

Fig. 2.5   Energy flow through the trophic levels (photo credit: S. Mantle)

In a food chain, "food" contains the base group of marticipating in a food chain. A food chain describes the transfer of energy and matter from primary producers to consumers to decomposers via a series of organisms that eat and are eaten. Various food chains come together to form a food web (Fig. 2.4). Each organism in an ecosystem can be assigned to a trophic level within its food chain or food web; each trophic level contains a certain amount of biomass (Fig. 2.5). That total biomass decreases as we travel up the food chain. The transfer of energy between these levels has a certain ecological efficiency. Energy is dissipated as it moves up each trophic level due to respiration and inefficient transfer of energy. The ratio is generally referred to as a law of 1/10; i.e. only 10% of the energy is captured at the next trophic level. This is a rough rule of thumb since some agricultural systems are actually considerably more efficient; e.g. broiler chickens.

2.4 Agroecosystems Harvest the Sun

One can sum up the energy relationships of an agroecosystem in a simple phrase: "All flesh is grass." Simply put, all ecosystems on earth are powered by the sun and its energy that is captured through photosynthesis. The role of agricultural production systems is to "harvest the sun"; i.e. as efficiently as possible capture as much energy per unit area to provide the power for the functioning of the ecosystem. Through generations of selection, breeding and more selection, our domesticated plants have been bred for increased efficiency at the cellular level in terms of photosynthesis. Domesticated plants are far more efficient than their wild predecessors in capturing energy from the sun and storing it in a usable packet for the humans that culture them. We not only have inherently more efficient photosynthetic factories in our crop plants but also have learned how to maximize their ability to intercept light through management practices. In short, the interaction of proper breeding and proper management of our crops means that we can capture high amounts of energy per hectare (ha, hm^2) of land. This is a crucial factor for overall agroecosystem efficiency.

Leaf Area Index (LAI) and Leaf Area Duration (LAD)

A very simple rule of thumb to be used in maximizing light interception by a crop is that the farmer should manage their crop in order to allow it to establish the optimum leaf area index (LAI) as quickly as possible and maintain it throughout the duration of the available growing season. LAI is the ratio of the area of green leaf per area of land; e.g. an LAI of 3 means that there are 3 m^2 of leaf per m^2 of land. Approximately 90%–95% of the light is intercepted when the leaf area index is 3.0–3.5. Increasing the leaf area index beyond that would be uneconomical. As it turns out, in many cropping situations, the optimum leaf area index is in the range of 3–4. The optimum leaf area index is one in which most of the light is intercepted and there is little apparent bare ground (Fig. 2.6). At this point, the crop is efficiently intercepting a majority of the energy from the sun.

Once the crop is established with an optimum leaf area index, in order to obtain maximum crop production, it is necessary to maintain that healthy green leaf area index for as long as possible

Fig. 2.6   An example of a soybean field where optimum leaf area index has been achieved and there is little apparent bare ground (photo credit: pixabay.com)

within the growing season. This introduces the concept of leaf area duration (LAD). This is essentially the product of time x leaf area index, or how long the optimum leaf area index is maintained. The important thing to remember here is that crop yield is directly proportional to LAD! Therefore, for maximum yield, the producer must manage their crop to get the optimum leaf area index as soon as possible and keep it there as long as possible in order to have the most energy captured from the sun. From an ecological standpoint, this is very important since optimizing yield on a per land area basis will minimize the amount of land required to produce the food that is needed. When a farmer is inefficient and produces low yields on good land, it means that we end up farming on marginal land that is more fragile, and the agroecosystem displaces natural ecosystems needed for global balance and other ecosystem services beyond providing food.

2.5   Matter Cycle and Agroecosystems

Through complex sets of interconnected cycles, micro- and macronutrients circulate within the ecosystems where they are most often bound in organic matter. The efficiency of nutrients entering and cycling through an ecosystem is based on hydro-geochemical processes, biological components and the maturity of the system. A natural ecosystem tends to regulate nutrients by recycling and replacing losses with local inputs. Nutrients are continuously cycled in nutrient cycles or biogeochemical cycles. The hydrologic cycle collects, purifies and distributes Earth's water. Other examples of cycles include the carbon cycle, and nitrogen cycle, phosphorus cycle and sulphur cycle. In an agroecosystem, human activities alter the cycles to regulate or enhance or improve nutrient balances, sometimes inadvertently causing severe disruption of the cycles themselves.

Biological components of each system become very important in determining nutrient movement efficiency, ensuring that a minimum amount of nutrients is lost in the system. Annual nutrient loss can be devastating from harvesting, leaching and erosion due to improper practices. Agroecosystems are inherently nutrient leaky systems that need to be carefully managed to keep the losses to a minimum. Nutrients can be leaked from an ecosystem from exposed soil between crop plants and during periods of bare ground between cropping cycles. Modern agriculture relies on significant external inputs to replace the lost nutrients and sustain the system so it can continue to produce. These external inputs are necessary to balance against the outputs of sold material, such as vegetables, milk and meat, but the unintended losses of nutrients should be kept to a minimum both for economic and environmental reasons.

Fertilizers, pesticides and machinery-based inputs maintain the economic viability of agroecosystems, increasing the total carbon footprint of the system. In more developed countries, increased input results in more crop yield per ha. For example, Japan and the USA have both the highest fertilizer and additional inputs and the greatest crop yield per ha in the world. In contrast, Indian agriculture represents the lowest input pattern in the world. These yield results encourage countries to increase the man-made inputs to respond to the food production pressure of growing populations. However, tremendous problems can accompany these crop production methods. Therefore, more targeted inputs, higher efficiency can be the key to managing sustainable agroecosystems. In summary, present high input farming practices are energy and matter intensive. There is a need for reducing man-made sources of energy to balance natural nutrients in food production.

2.6　Ecological Footprint

The concept of ecological footprint, which involves measuring human impacts on land and water use, resource consumption and production of wastes, has been used originally in helping communities look at alternative ways of resource management. More recently, the tool is being used for the education of individuals and groups on our impact on the world.

Today, humanity's ecological footprint is larger than what the planet can regenerate. When we consider our individual ecological footprint, we gain perspective on both our personal impact and how our actions as a collective can support the goal of humanity living within the means of our one planet. In this textbook, we are focused mostly on the components of the ecological footprint linked to agriculture. To do that we need to consider four questions:

1. What are the resources that are used in growing crops and raising animals and taking products from soil to shelf?
2. What is required to dispose of wastes?

3. How can our agricultural footprint be decreased?

4. The definition uses the term "prevailing technology"; is new technology the answer?

The reader is urged to keep these four questions in mind as we look at how our production process can change to lower our ecological footprint in response to agriculture.

Part II
Basics of Agroecosystems

Chapter 3
Soils as the Basis for Cultivated Ecosystems

D. H. Lynch and S. Smukler

"Land is a blessing and a gift from God and is, therefore, sacred. It is the source of life of the people like a mother that nurtures her child. Consequently…land is life."
—Ponciano L. Bennangen

Abstract Intensive farming systems are a major driver of land degradation and soil losses and declines in the abundance and diversity of animals and plants. Through an improved understanding of soils and soil health and the impacts of agroecosystem management on soils and soil functioning, we can develop and support farming systems that are sustainable both ecologically and economically while also providing sufficient supplies of food and fiber.

Learning Objectives
After considering this topic and reading this chapter, students should be able to:

1. Explain the origin of soils, and the factors that determine soil formation.
2. Describe soil horizons, and soil classification, and give examples of soil orders.
3. List the ecosystem services of soils.
4. Explain how soil texture and soil structure affect soil function.
5. Explain the importance of soil organic matter in agroecosystems.
6. Define the following terms:
   – Soil cation exchange capacity.
   – Soil pH (acidity and alkalinity).
   – Salinity of soil:

   (a) Soil erosion
   (b) Soil compaction
   (c) Soil degradation

7. Describe a general soil nutrient cycle.
8. Explain the role of soil ecology in sustainable management of agroecosystems.
9. Explain the terms soil quality and soil health and describe their impact on the provision of ecosystem services.

10. Provide examples of best management practices in temperate and tropical agroecosystems for maintaining soil health and plant nutrient supply, and mitigating climate change.

3.1　Introduction

A soil is a complex dynamic matrix that includes both abiotic components derived from rocks and biotic components, living organisms. Soils are evolving, dynamic complex ecosystems that have played an important role in human history and will inevitably do so in our future (Lynch, 2019a). Soils are a key component for producing food, fuel, fiber, and timber. But soils do much more than the critical role of supporting plant growth; fully functioning soils, whether in natural or managed terrestrial ecosystems, are the central interface (as the pedosphere) between the Lithosphere, Biosphere, Hydrosphere, and Atmosphere, linking the global carbon, nutrient and water cycles. We are increasingly recognizing soils as far more than solely an anchor for plants, but as the irreplaceable "skin of the Earth" providing, within all agroecosystems, economic, environmental, and social services that are essential for life (Lynch, 2019a). Thus, preserving soils and soil quality is critical to food security, preserving biodiversity, and tackling climate change while enhancing agroecosystem resilience to the stresses imposed by human activities and the volatile weather induced by a rapidly changing climate.

3.2　Formation and Classification of Soils

A soil can be defined as any naturally occurring, unconsolidated (loose) organic and mineral material on the surface of the earth, which will support plant growth (NRC n. d.). Recent revisions to this definition suggest it is important to acknowledge that soils also hold liquids, gases, and diverse organisms, and need not always support plants (for example in deserts) (SSSA n.d.). Through the process of soil formation (soil genesis), soils are formed over many thousands of years from their parent materials. The intensity of these soil-forming processes vary with the local environment. Thus, in practical terms, soils are considered a nonrenewable resource.

Soils are a product of their Parent material, plus Climate, Topography, Organisms (including Humans), and Time.

Parent Materials The parent material is the original mineral material, or less often the organic material, from which soils form. It can be comprised of mixed materials (sand, silt, clay, and gravel) deposited by glaciers or sorted by water or wind during or after glacial retreat.

Climate Including temperature and precipitation, climate is a major driver and determinant of soil genesis. Through its influence on the dominant vegetation, biological processes in soil including mineralization, and in regions of higher precipitation, leaching of soil nutrients (e.g., cations), organic matter, and clays into lower layers (horizons) of the soil profile.

Topography The shape and position of the soil in the landscape, through its influence on soil moisture and water movement, influences soil development.

Organisms Soil organisms, plants, and animals (including humans) all influence soil genesis. Soil formed under grasslands for example tend to have the highest soil organic matter content, forest soils intermediate, and desert soils the lowest soil organic matter.

Time The processes of soil genesis, influenced by biological, chemical, and physical processes determined by the above factors, takes many thousands of years.

Soil horizons are the identifiable and distinct layers in the soil profile based on the composition of the layer and its color, structure/texture, or chemical properties, that develop over time through the processes of soil genesis, as influenced by the local environment. The vertical arrangement of soil horizons comprises the soil profile and is used to distinguish and classify soils. Fig. 3.1 shows a typical soil horizon.

Soils are classified globally, and by country, in a hierarchical system in which soil order is the highest rank and local soil series the lowest. In Canada, for example, there are 12 distinct soil orders (www.soilsofcanada.ca).

3.3   Ecosystem Services Provided by Soils

Despite the key central role of soil and a multitude of soil functions, in terrestrial ecosystems, including agroecosystems, the contribution of soils to human welfare beyond food production, is still underappreciated. Recent research is increasingly documenting the contribution and relationship of soils to the four categories of ecosystem services, namely, Provisioning (food, fresh water, wood, fiber, and fuel), Regulating (regulation of air and water, climate, floods, erosion, and biological processes such as pollination and diseases), Cultural (esthetic, spiritual, educational, and recreational), and Supporting (nutrient cycling, production, habitat, biodiversity) services (Adhikari and Hartemink, 2016).

The FAO (Fig. 3.2; FAO, 2015a) lists 11 ecosystem services associated with soil.

Fig. 3.1　Picture of a typical soil horizon showing distinct layers in the profile (photo credit: G. Brewster)

G.　Brewster

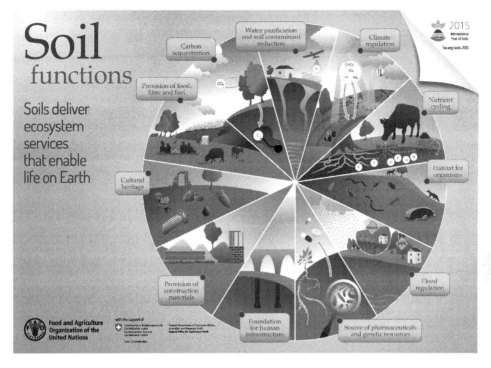

Fig. 3.2 Soil functions (FAO, 2015a)

3.4 Soil Characteristics

3.4.1 Soil Physical Properties

Soil is a three-dimensional matrix comprised of open spaces (pores) for air and water, organic matter, and the solid sand, silt and clay particles. The proportion of sand, silt, and clay particles in a soil is referred to as *soil texture*, and we can classify soils according to their textural class of which there are 12 classes (see Fig. 3.3). Soil texture is generally considered an inherent or intrinsic property of soil, not practically or readily modified by management practices.

Note that clay is the most influential of the soil particles as it has the most dominant influence on soil properties. This is because many of the key physical and chemical reactions of soil occur on the surface of soil particles, and the specific surface area of the particles increases as the size of the particle decreases. Clay particles, which are smaller than 0.002 mm, are often shaped likes platelets, thus have a relatively very large specific surface area. Thus, soils with higher clay content have a greater ability to retain some nutrients (see CEC in sect. 3.4.3) and also form chemical organo-mineral complexes that help retain soil organic matter. Textural classes (Fig. 3.3) are defined based on the relative sand, silt, and clay composition, ranging from coarse-textured soils dominated by sand to finer textured soil more dominated by silts and clays, with loams as intermediate.

Soil structure refers to the three-dimensional arrangement of the solid phase of the soil, the soil particles, and the soil *pore* spaces. Soil particles, along with soil organic matter, clump together to form *aggregates* of various sizes. These aggregates, and the complex arrangement of micro- and macropores between these aggregates, comprise the soil structure. Ideally, the pores spaces in soil should be close to 50% of the total soil volume, with the mineral fraction (sand, silt, and clay combined in aggregates) often 45% and organic matter typically up to 5%. The biotic component of soil, living plant roots, and soil organisms (especially during the process of decomposition of organic matter) play a key role in the formation of soil aggregates and maintenance of soil structure.

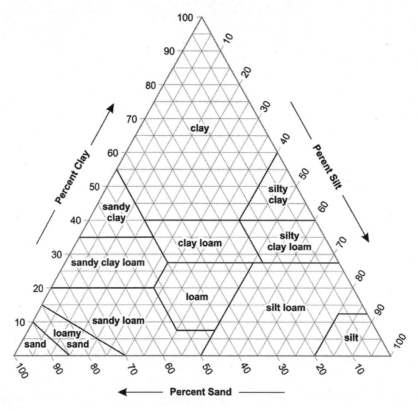

Fig. 3.3　Soil texture triangle showing the relationship among concentrations of sand, silt, and clay and 12 texture classes (NRCS/USDA)

Together soil structure determines how effectively plant roots can penetrate the soil enabling plant growth, provides the habitat for soil organisms, allows air and water movement into and through the soil, provides for the water holding capacity of the soil, and allows the critical soil functions including soil organic matter decomposition and recycling, and nutrient cycles, to occur. Unlike soil texture, soil structure is a dynamic property of soil, very sensitive to farming intensity and management practices. Thus, maintaining the soil structure is key to sustaining the optimum functioning of the soil and its resilience to stresses.

3.4.2   Soil Organic Matter and Its Importance in Agroecosystems

Soil organic matter is decomposing material derived from previously living plants and organisms. It is a very diverse material in soils, from more recently deposited plant residues or litter to highly decomposed amorphous humus, combined with the by-products of the microbially driven process of decomposition. Soil organic matter also includes the biotic component (living organisms) in soil. Soil organic matter is rich in carbon (containing approximately 50% carbon) and soil organic carbon (SOC) is the keystone element linked to soil quality and soil health (Lynch, 2015). As most soil organisms are heterotrophs, SOC levels and the cycling of SOC in the soil is critical to providing both habitat and energy for these organisms.

Soil organic matter plays a central role in maintaining soil structure and soil health and ensuring soil functions optimally—through its dominant influence on soil physical, chemical and biological properties and processes. Soil organic matter is critical to:

Soil Structure Aggregation, porosity (water infiltration and storage, airflow), soil tilth, and optimum conditions for microbial processes, habitat for soil biota.

Soil Biological Properties Energy and nutrient source for soil biota.

Soil Chemical Properties Holds nutrients (exchangeable cations on CEC (see sect. 3.4.3) and is an important source of others (N, P, S).

Thus, while soil organic matter typically comprises between 1% and 5% of the mass of most mineral soils (the remainder being sand, silt, and clay particles), its influence on soil physical–chemical and biological properties and soil functions, far outweighs its relative contribution to the soil mass. Declining soil organic matter levels and associated losses in soil quality and soil health and productivity are a critical issue globally (Lynch, 2019b).

Maintaining soil organic matter is also critical to much needed nature-based solutions to climate change (see Chap. 10 on Climate Change). We are increasingly appreciating that soils are key to turning back the carbon clock and reversing atmospheric CO_2 accumulation (Lynch, 2019a) Soil carbon is also the keystone element controlling soil health, which enables soils to be resilient to increasing droughts and intense rainfall events expected as a result of climate chaos (Lynch, 2019b).

The total SOC (and soil organic matter) stored in a soil is determined by the balance between the processes of decomposition of C from crop residues and organic amendments, and the losses of C as a result of soil respiration, which is enhanced by tillage and other soil disturbances. Carbon is continuously flowing through the soil ecosystem and the total SOC "stored" in the soil profile is in a state of dynamic equilibrium with constant C additions (necessary to promote soil health) balanced by losses. When this balance is upset through a change in ecosystem or soil management (resulting in reduced C additions or additional SOC losses from soil), the system transitions to a new equilibrium level of reduced SOC storage (Fig. 3.4).

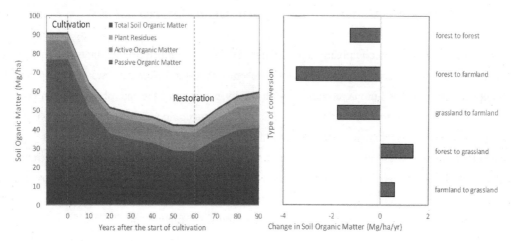

Fig. 3.4　Expected pattern of soil organic matter dynamics after the start of cultivation for active and passive pools (left; modified from Weil and Brady, 2017) **and global averages of soil organic matter losses and gains following the conversion of forest, grassland, or agriculture** (right; modified from Deng et al., 2016)

In North America, it is estimated that agricultural soils have historically lost up to 50% of their original native SOC levels. In Canada, farm SOC levels have stabilized over the past few decades on the Canadian Prairies as a result of the adoption of no-till cropping systems. In contrast, in the more humid regions of Eastern Canada and British Columbia, the increased intensity of cropping (especially reduced use of perennial crops), and lack of SOC gains from no-till systems, is leading to declining SOC levels across the farms of the region (Lynch, 2019b).

3.4.3　Soil CEC, Acidity and Alkalinity, and Salinity

Cation Exchange Capacity (CEC) The CEC of a soil is the capacity of the soil to hold, by adsorption, exchangeable cations (positively charged molecules or ions). Fine particle (colloids) of clay and humus (very old organic matter) in soil have a large relative surface area and negative charge on their surfaces and attract cations (positively charged ions) some of which are nutrients, including potassium, calcium, magnesium, and ammonium-nitrogen. The total CEC is the negative charge per mass of soil (meq/100 g). The CEC thus describes the potential nutrient holding capacity of the soil. Building the organic matter content of a soil increases its CEC.

Soil pH (Acidity and Alkalinity) Soil pH is a measure of the concentration of H^+ ions in the soil solution: $pH = -\log [H^+]$. The higher the H^+ concentration the more acidic the soil. pH is represented across a logarithmic scale ranging from 0 to 14, with a pH of 7 being neutral, soils lower than 7 acidic, and those above 7 alkaline.

Both soil biological activity and soil plant nutrient availability are strongly influenced by the acidity or alkalinity of a soil, with both tending to be greater or optimized in near neutral (~pH 6.5) soils. Also, crops vary in their preferred soil pH with most agricultural crops, with some exceptions such as blueberries, preferring pH in the range of 5–8. Given the improved productivity of crops at or just above a pH of 6.5, farmers generally add lime to their soil to raise or maintain the pH. Without lime, agricultural soils typically acidify with the addition of fertilizers.

Salinity In arid and semiarid regions with limited rainfall and high evaporation, soluble salts are not leached down through the profile and naturally accumulate in the upper soil layers, resulting in a saline soil. This may be exacerbated by increased intensity of cropping (including the addition of synthetic fertilizers and the use of irrigation). High levels of soil salinity [measured as electrical conductivity, EC (ds/m)] can limit the cropping options and also negatively affect soil physical properties.

3.4.4 Soil Nutrient Cycles

Nutrients, like carbon (or energy), are cycled within an ecosystem. A natural soil ecosystem can function with no additions of nutrients almost indefinitely, unless there is some net loss of nutrients from the system. Most natural soil ecosystems tend toward an equilibrium state wherein the losses are balanced by natural inputs. The nutrients cycle within the system and constantly change by natural processes from one form to another. It is this continual process of biological–geological–chemical transformations, that we call the Soil Nutrient Cycle.

Soils in managed agroecosystems differ from unharvested soils in that ①the nutrient cycle is usually intensified and ②nutrients are removed with the harvested crops, plus ③farm management practices usually increase the amount of nutrients lost from the system other than by crop removal such as through erosion of soil, and nutrient losses through leaching and gaseous loss (see Fig. 3.5). Depending on the farming system employed also, the reliance on inputs of synthetic nutrients can vary substantially. Organic farming systems, for example, rely more on soil biological nitrogen fixation by legumes to provide N to the systems, and/or soil mineralization to make organic N available to the crop (Woodley et al., 2014).

The macronutrients, those required as *essential plant nutrients* in the greatest amount by crop plants, include nitrogen, phosphorus, potassium, calcium, magnesium, and sulfur. In addition, plant essential micronutrients like boron, zinc, manganese among others, although required in lesser amounts by plants, can also greatly limit crop growth if insufficiently supplied by soil and/or nutrient inputs in agroecosystems. Plant productivity is thus not necessarily determined by the overall quantity of nutrients present in the soil but by the scarcest nutrient in need. This concept is often referred to as Liebig's Law of the Minimum.While plant growth can be increased by the addition of nutrients, growth response is typically the greatest with lower native availability of soil nutrients, declines as the nutrient availability increases and eventually reaches a maximum after which greater nutrient availability does not result in increased productivity.

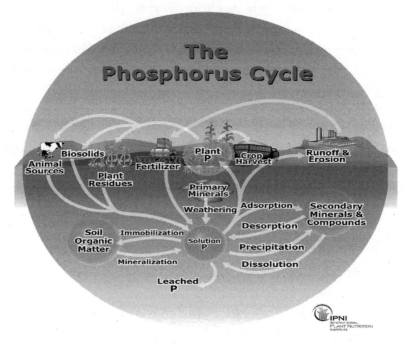

Fig. 3.5　The soil phosphorus cycle (IPNI)

The stores of nutrients in soils (Fig. 3.5) include:

Living Biomass Nutrients contained within plants and soil biota.

Soil Organic Matter A significant source of nitrogen, phosphorus, and sulfur.

Inorganic Store Soil minerals (rocks and minerals) slowly releasing nutrients plus exchangeable cations (including potassium, calcium, and magnesium) adsorbed on the CEC.

Various soil chemical and biological *transformation* processes enable a fraction of these stored nutrients to enter into the soil solution and become plant available. For example, the biological conversion of the organic forms of nitrogen, phosphorus, and sulfur into plant-available forms is called *mineralization*. Plant roots take up the nutrients from the soil solution in a process called *absorption*.

Gaseous nutrient losses from soil and agroecosystems include nitrogen lost as ammonia (NH_3–N) and nitrous oxide (N_2O). The latter is a very potent greenhouse gas, of which agriculture is the primary source globally. In areas of high precipitation, losses, through leaching and/or runoff of nitrogen and phosphorus is a serious issue globally, not only because of the loss of nutrients but because of the negative off-farm environmental impact of these reactive nutrients on water quality and aquatic systems including lakes and inshore fisheries. Nitrate nitrogen also can lead to the degradation of community drinking water quality.

3.4.5   *Soil Ecology and Its Role in Sustainable Management of Agroecosystems*

Soil contains one-quarter of all the world's biodiversity; it is where many plants, bacteria, and fungi evolved together. In many cases, plants and soil microbes established mutually beneficial relationships, communicating with each other by sending chemical signals through the soil (Lynch, 2019a). The porous structure of soil, and variable supply of organic residues, food, water, and chemicals, provides a range of habitats and niches for a multitude of macroorganisms (>2 mm) to microorganisms depending on climate, vegetation, and soil properties. A teaspoon of soil may contain hundreds of nematodes, and thousands of algae, amoeba, fungi, actinomycetes, and bacteria (Plaster, 2009). Recent advances with molecular techniques are greatly improving our understanding of the diversity of, and linkages between, soil biota, i.e., the diverse micro- and macroorganisms in soil (Atlas of soil biodiversity). Soil scientists are also increasingly examining how the essential functions of soil organisms respond to stresses from human activities and a changing climate.

The essential functions performed by soil organisms include:

- Nutrient cycling
- Residue decomposition and soil carbon sequestration
- Greenhouse gas emissions
- Modifying and maintaining soil structure and soil water regimes

Maintaining the biological characteristics of soil and overall soil health, as both play a key role in the processes of decomposition and nutrient cycling and sustain soil resilience to stresses including those derived from climate change and human activity, is essential to sustainable soil management.

The interactions in the soil ecosystem between soil organisms and between soil organisms and plants and animals in the ecosystem is called the soil food web (FAO, 2015b). Plants are the *primary producers* (and comprise the first trophic level or food base in the ecosystem), as they produce their own food (i.e., autotrophs producing living organic materials from CO_2 and sunlight). The *primary consumers* are those soil organisms (microbes and macro- and mesofauna) that consume the energy-rich living tissue or decaying residues from plants. In turn, and at the next trophic level in the system, predators are *secondary consumers* that feed on the bodies of primary consumers to derive energy and nutrients (Plaster, 2009). *Decomposers* are the essential organisms in the soil that close the nutrient cycles. Decomposers break down organic materials and provide nutrients for plant growth, releasing CO_2 from waste while consuming O_2 produced by photosynthesis. Decomposers include various types of insects, fungi, bacteria, and earthworms.

More intensive cropping systems (reduced cropping diversity and short rotations, increased tillage or soil disturbance, low residue, or organic matter return to soil) can degrade both soil structure and also disrupt the microbial and meso-faunal communities in soil. Soil organisms with larger body size, including earthworms, nematodes, and microarthropods are more sensitive to agricultural intensification than soil microbes generally (Postma-Blaauw, et al. 2010; Lynch, 2014). The high levels of tillage and soil disturbance, and low residue return to the soil, commonly associated with some crops, such as potatoes, are also deleterious to soil life. However improved cropping management regimes can mitigate against these impacts. In Atlantic Canada, the longer (5 year) crop rotations on organic potato farms, including 2–3 years of perennial forages, allowed earthworm and soil microbial populations to recover from being greatly reduced during the potato crop year (Nelson et al., 2009). Mann et al. (2019) assessed soil health and soil microbial diversity on 34 diverse farms in Atlantic Canada. Lower-intensity management and cropping practices (i.e., use of perennial forages, and/or mixed annual–perennial cropping), along with manure application and less tillage were found to be associated with increased soil fungi, mycorrhizae, and Gram-negative bacteria, along with improved overall soil health.

3.4.6　Soil Quality and Soil Health

Given that soils are a critical component for many ecosystem services, it is essential to have a way to describe, evaluate, and even quantify their capacity to provide these services. To do this, scientists and land managers have developed the concepts of soil quality and soil health. Since first being introduced in the 1990s, these concepts have evolved and there has been great debate as to the best ways to evaluate them and even how to define them. There are those who think that the terms soil quality and soil health can be used interchangeably (Bünemann et al., 2018) while others consider these terms completely distinct (Lal, 2016).

Soil health is defined as: "the continued capacity of soil to function as a vital living system, within ecosystem and land-use boundaries, to sustain biological productivity, promote the quality of air and water environments, and maintain plant, animal, and human health." (Pankhurst et al., 1997)

Soil quality, while similar in concept has been distinguished from soil health as being specific to functions for a particular land use. Soil quality relates to the ability of a soil to carry out the major functions of soil as noted in Fig. 3.2, specifically the capacity within either natural or managed ecosystems to support plant and animal production, maintain or improve water and air quality, and provide support to human activities.

Both soil health and soil quality are challenging to measure directly and there have been a number of proposed methods for doing so.

Soil Health/Quality Indicator A measurable soil property, either physical, biological, or chemical or a soil function that is used as an indirect proxy for evaluating soil health or soil quality.

Soil Degradation Loss of soil quality resulting from human activities or management resulting in a reduction in the soil's ability to provide ecosystem services.

Erosion Soil movement and loss by wind, water, or tillage.

Compaction Physical forces applied to soil resulting in soil particles being squeezed more tightly together, reducing pore space for water and air, and restricting root growth.

Independent of whether a soil quality or soil health framework is adopted to assess the sustainability of farming systems, it is important to examine specific farming practices or other land management regimes and avoid soil degradation; both concepts and frameworks are comprised of integrated indices of a soils ability to function optimally. As integrated frameworks of soil condition, they analyze (in situ or in laboratory) and assess some combination of soil physical, chemical, biological and biochemical properties. This assessment framework is ideally regionally adapted to suit the local edaphic and agricultural conditions. In recent years in North America, the agricultural sector and government-led programs to promote soil health and develop regionally relevant testing soil health protocols have become widely adopted. The USA-based Soil Health Institute (https://soilhealthinstitute.org) is an example of a lead organization coordinating some of these efforts.

As described in Sect. 3.4.5, less intensive management systems (reduced tillage, extended and more complex rotations, increased use of forages and cover crops) can increase SOC levels and provide the co-benefit of enhancing soil health. In tropical systems, slash and mulch agroforestry show promise to reduce the need to rotate crops into forested regions, while maintaining soil health and ecosystem services (see Fig. 3.6 and Box 3.1).

Fig. 3.6   Maize and beans growing among nitrogen-fixing trees on a steep slope in El Salvador as part of a slash and mulch agroforestry system (photo credit: S. Kearney)

Box 3.1　Slash and Mulch Agroforestry in the Tropics

In many parts of the world, indigent farmers are forced to grow crops on small tracks of land often in areas that are marginal for production because of poor soils or steep topography. In some parts of the tropics, "smallholder" farmers have maintained agricultural productivity despite these conditions, by cutting down patches of forest and burning them to release the nutrients for their crops. These nutrients will often be enough to supply their crops for a few years but eventually, crop yields decline as does the quality of the soil. Slashing and burning typically happens just before seasonal rains, and without forest cover protecting the soil, the soil erodes rapidly. Once crop yields decline these farmers will then either let the land revegetate naturally or graze the land with their animals until that too becomes unproductive. The farmers will leave the land fallow for years until the fast-growing tropical trees are large enough to start the cycle again. While this type of slash and burn agriculture was sustainable when there was an abundance of forest, because of rapidly expanding human populations, this is no longer the case. There is now not enough land to allow for fallows of a duration to accumulate the nutrients necessary to sustain crop productivity. There is, however, a promising alternative to slash and burn that has been practiced traditionally by smallholder farmers in the mountains of Honduras called Quesengual or slash and mulch agroforestry (Hellin et al., 1999).

Slash and mulch agroforestry is designed to protect the soil and build soil organic matter and thus maintain a slow and steady release of nutrients to the crop over time. Instead of cutting down all the trees in an area being prepared for cropping, a few key trees are left standing and are pruned. The wood from the trees that have been cut is then sold as firewood and the leaves and branches are left on the soil surface as a mulch to decompose. The mulch protects the soil from intense tropical rainfall and the rapid decomposition in the warm, wet tropical climate provides nutrients to soil organisms and ultimately to the crop. The trees left standing in the field help stabilize steep soils with their roots, further reducing erosion. Farmers generally select trees that are nitrogen-fixing thus the trees act as a source of nitrogen as well. Before the crop is planted each season, the trees in the crop field are again pruned and their nitrogen-rich leaves used as mulch and fertilizer. In the mountainous region of nearby El Salvador, researchers have trialed another type of slash and mulch agroforestry where large tree stakes are transplanted into fully cleared fields in order to establish an agroforestry system in areas that had already experienced slash and burn (Kearney et al., 2017). After 3 years, researchers found that crops grown after the traditional slash and mulch agroforestry originating from the forest had greater mulch cover, lower rates of runoff, and higher rates of water percolation than crops following slash and burn but they also had lower yields of corn than slash and burn. Alternatively, corn yields in the new agroforestry slash and mulch had the same yields as the slash and burn fields. Both agroforestry systems had lower weed pressure and higher numbers of earthworms than the slash and burn. These two slash and mulch agroforestry options, by maintaining soil health and function show promise to reduce the need for rotating crops into forested areas and to sustain a greater amount of ecosystem services in a region.

Conclusions

Intensive farming systems are a major driver of land degradation and soil losses, and declines in the abundance and diversity of animals and plants (https://www.ipbes. net). Through an improved understanding of soils and soil health, and the impacts of agroecosystem management on soils and soil functioning, we can develop and support farming systems that are sustainable ecologically while also providing sufficient supplies of food and fiber.

Chapter 4
Water as the Basis for Cultivated Ecosystems

S. Wang and C. D. Caldwell

> *"We're all downstream—Ecologists'motto, adopted by*
> *Margaret and Jim Drescher, Windhorse Farm, New Germany,*
> *Nova Scotia, Canada."*
> — *Marq de Villiers*

Abstract Water is the main constituent of all living organisms including humans; it is estimated that the human body contains 70% water. Water is a basic requirement for successful agricultural production to feed the world's population. Water availability is the key limiting factor among all factors that determine crop yield. Even though the earth is named a "water planet" that uniquely accommodates life, the usable water for agriculture is extremely limited. In addition, due to a lack of knowledge and awareness, we have abused the water both qualitatively and quantitatively leading to widespread soil water deficits and water contamination worldwide. Knowledge and action on water-related issues, especially as they relate to agroecosystems, are key to our future.

Learning Objectives

After studying this topic, students should be able to:

1. Explain the significance of water for people in relation to agroecosystems.
2. Describe the hydrologic cycle.
3. Describe how human activities are affecting the hydrologic cycle; use two examples of issues concerning modern agriculture and water to illustrate these impacts.
4. Explain the social and economic impacts of these activities.
5. Define and describe the major aspects of managing water in agroecosystems.

4.1 Water Resources in the Global Ecosystem

Water covers about 71% of the Earth's surface; however, only 2.5% of the water on the Earth is freshwater. Most of the freshwater on the Earth is locked in glaciers or too deep in the ground to be used, leaving less than 1% of water on the Earth available for human consumption (Grey et al., 2013). Important freshwater resources, such as groundwater, would diminish within a short time period without replacement via the hydrologic (water) cycle (Table 4.1). The hydrologic cycle refers to the continuous movement of water throughout the Earth, purifying, recycling, and replenishing reservoirs (Fig. 4.1). Groundwater is often used for agricultural, commercial, and industrial and personal uses when the water flows to the surface naturally forming springs, seeps, oases, and wetlands (or into man-made wells).

Table 4.1　Water in the global ecosystem (Dyck and Peschke, 1995)

Reservoir	Area (km²)	Volume (km³)	% of all
Oceans	361,300,000	1,338,000,000	96.5000
Groundwater	134,800,000	23,400,000	1.7000
Soil water	82,000,000	16,500	0.0010
Ice and snow	16,227,500	24,364,100	1.7700
Freshwater lakes	1,236,400	91,000	0.0070
Salt water lakes	822,300	85,400	0.0060
Swamps	2,682,600	11,470	0.0008
River water		2120	0.0002
Water in biota		1120	0.0001
Water in atmosphere		12,900	0.0400

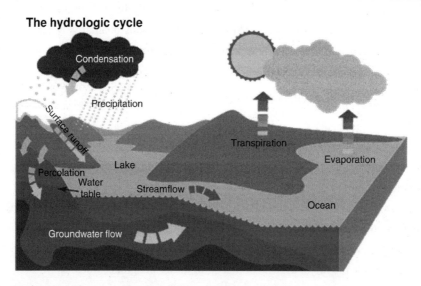

Fig. 4.1　The global water cycle (https://www.canada.ca/en/environment-climate-change/services/water-overview/basics/hydrologic-cycle.html)

Freshwater is the lifeblood of the Earth and forms the human drinking water supply. However, due to human abuse and contamination of our waterways, provision of safe drinking water has become a challenge worldwide; approximately 1.2 billion people do not have access to safe drinking water and over 10 million people dwelling in mega-cities throughout the developing world heavily depend on groundwater (Vörösmarty et al., 2015). While 500 million people live in water-scarce and water-stressed countries, even in locations where water is plentiful, many poor people cannot afford a safe supply of drinking water. According to the UN agenda for the third millennium, the deficit of fresh water for human society reached 230 billion m^3 in the year of 2000, and the number will increase to 1.3–2.0 trillion m^3/year by 2025. Moreover, frequent occurrences of prolonged droughts caused by global climate change and water scarcity by human abuse kill more than 24,000 people a year and have created millions of refugees since the 1970s.

The three main pressing water-related problems confronted by humans today are:

1. Groundwater overexploitation. Over pursuit of economic development in the last decades both in developed and developing countries with expanse of more freshwater input usually led to overuse of groundwater, which in turn has caused water table imbalances, ecological and health issues. For example, to produce electrical power and reduce flooding, we build dams with powerful technology; these dams often increase the annual runoff available for human use, but also reduce downstream flow drastically and can prevent rivers from reaching the sea, thereby radically changing the region's entire ecosystem. Transferring water through tunnels, aqueducts, and underground pipes often draw groundwater faster than it can be renewed. These practices change the water table levels and create serious ecological and health disturbances in the ecosystem.

2. Water pollution. Worldwide water pollution occurs both on the surface and underground due to contamination by industrial wastes and overuse of agricultural chemicals. Water pollution refers to a physical, chemical, or biological change, from a point or nonpoint source pollutant that reduces water quality. Some major categories of water pollutants include oxygen-demanding wastes, inorganic chemicals, organic chemicals, plant nutrients such as nitrogen and phosphorus, radioactive material, and heat. Measuring biological oxygen demand or taking a chemical analysis can indicate pollutants from point sources. Once identified, point sources can be monitored and regulated. Unfortunately, it is more difficult to control nonpoint source pollutants. Lakes and ponds are quite vulnerable to pollution because they have little flow. However, flowing bodies of water such as streams may rapidly recover from pollution through dilution and bacterial decay, so long as they are not inundated with contaminants. Laws concerning water-pollution control have significantly improved stream water quality in most developed countries, but political infrastructure in developing countries has not been able to enforce pollution control and stream water quality remains poor in those places. For example, in China, there is a serious water pollution issue with the paper-industry dumping toxic chemicals and waste in rivers. In 2004, one-third of factories lining the Huai River, which supplies one-sixth of the nation with drinking water, still did not meet waste dumping government standards. Many areas such as the Huai Lakes in southern China are facing this problem (Liu and Diamond, 2005; Shao et al., 2006).

3. Inefficient water use (causing water cycle imbalance). Using water inefficiently is another major area of concern and only amplifies the problems to overexploit the land for more water. In the United States, the world's largest water user, nearly half of the water it withdraws is lost in some way. In China, especially in the west of the country, irrigation systems are unchanged from what was used 6000 years ago for a much smaller population.

Solving these problems will require an integrated approach to fixing each of these issue areas. The first step will require using water more efficiently. Redesigning national wateruse policies will provide incentives for water conservation and higher efficiency. Exploring more efficient irrigation system technologies for use in the world's croplands will also conserve water long term. A blue revolution will initiate and encourage more sustainable water management in the future. Improving water quality will be an important facet of sustainable water management, preventing pollutants from reaching the surface or groundwater.

4.2　Water in Agroecosystems

Agriculture uses the most freshwater of any industry sector, generally for irrigation. In fact, irrigation for agricultural purposes accounts for 80% of worldwide freshwater consumption (Liu and Song, 2019). The need for irrigation is great; because water is often the limiting factor on production, worldwide; while irrigated farmland is only about one-sixth of the total farmland, it accounts for more than one-third of the global harvest (de Villiers, 2000). Since the Green Revolution at the beginning of 1960, in some developing countries, irrigation takes up to 90% of the national water withdrawn; while slightly less water is used for irrigation in developed countries, it is still a significant proportion; for example, 30% of freshwater usage is for irrigation in the United States (Watson et al., 2014).

Irrigated water in agriculture gets used for growing crops, weed control, frost protection, and chemical applications. Usually, irrigated water comes from a higher percentage of surface water; however, in the USA, groundwater withdrawals for irrigation have persistently risen since the 1950s. Worldwide agriculture accounts for the largest water withdrawals and can be considered the major reason for water scarcity when water supplies cannot satisfy all demands (Sinha and Hyung, 2008). But such practices cannot go on indefinitely; it damages the water ecosystem services including provision (e.g., drinking water after purification), regulation, and support services and the monetary cost of groundwater is increasing greatly. Most importantly, overexploitation of ground water unravels the hydrological cycle in agroecosystems, which does harm to agricultural production in short term and undermines the civilization in long term.

4.2.1　Water Cycle in Agroecosystems

Fig. 4.2 shows a very general conceptualization, embracing most possible inputs and outputs to the water flows in an agroecosystem. Maintaining a healthy agroecosystem requires a balance among fluxes and balancing fluxes is quite complex. Agroecosystems, considered as parts of watershed ecosystems, can selfregenerate water resources via micro-cycles, which include multiple landscapes and biodiversity aggregations. Consider Fig. 4.3 which shows a micro-cycle of water. In these micro-cycles, rainfall water ("Yellow Stream") represents the basic water resource, partially assimilated by plants, partially evaporated at the water surface forming the "Green Stream" (65% of total). This is potentially very highquality water without human interference, which is available for the next water cycle. The remaining water flow is called the "Blue Stream" (35% of total) and it flows to aquifers underground or runs off via the soil surface to rivers and the sea. This Blue Stream also has the possibility of intervention and exploitation by human activity outside the watershed (Figs. 4.3 and 4.4) (Rauba, 2017).

Water cycles in agroecosystems are changing due to global warming, groundwater overuse and contamination and field temperature increases. Global warming is changing the water cycle routes in all ecosystems, including agroecosystems. Increasing CO_2 concentrations has globally and regionally changed the rainfall distribution in time and space. Rainfall (yellow stream in Figs. 6.3

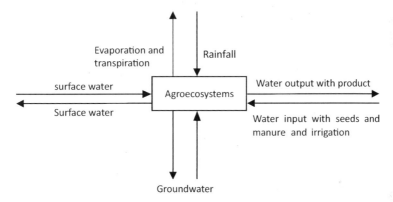

Fig. 4.2   The general pattern of water cycle in agroecosystems

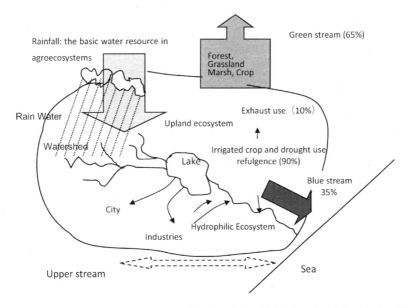

Fig. 4.3   Water cycle in agroecosystems (modified from Falkenmark and Rockström, 2006)

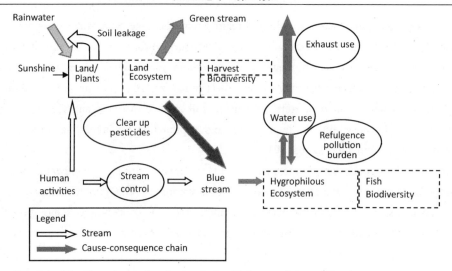

Fig. 4.4　The blue stream (surface water) with human intervention in agroecosystems

Human activities on the geomorphology change the blue stream directly by means of water controls in building and exhaust use of water, and indirectly by control of land and plants

and 6.4) has become more variable and erratic, causing a greater incidence of droughts and flooding. In the meantime, industrialization and urbanization, especially in developing nations, has increased the acidity of rainwater, destroying water quality. Another characteristically changing aspect of the water cycle is groundwater overuse and pollution. Groundwater has been overdrawn and heavily contaminated, decreasing its availability to agroecosystems. Finally, increasing field temperatures have led to higher evaporation rates, which increase water loss in agroecosystems (Fig. 6.4).

Meteorologists have mapped the global flow of huge amounts of water in the atmosphere; these are essentially flying rivers that transfer large amounts of water throughout the globe. Both global activities that have affected climate change and local activities of deforestation have had large effects on these flying rivers. We now know that transpiration moves more water than all the rivers in the world combined.

Forests throughout the world influence the atmospheric water cycle in various ways. These interactions are complex and require more research to properly characterize them to be able to predict the effects of deforestation both locally and globally. We do know that the destruction of forest cover affects the water cycle. Increasingly, agriculture threatens forests and other diverse vegetation as we change natural ecosystems into managed ecosystems for food production. The potential consequences of large-scale forest loss are severe. A recent theory regarding these interactions of vegetative change suggests that large changes in landscape transpiration can exert a major influence over atmospheric dynamics. This theory explains how high rainfall can be maintained within those continental land-masses that are sufficiently forested (Sheil, 2018).

A possible illustration on a micro-scale of this feedback mechanism may be playing out in Australia. Southwest Western Australia has experienced a decline in rainfall over the last 40 years. This is usually explained through natural variation and some effect of global warming; however, there is recent evidence to suggest that this decrease in rainfall may be substantially due to large-scale logging that has occurred close to the coastal areas. Models proposed by Andrich and Imberger (2013) show that between 55% and 60% of the decrease in rainfall is probably due to land clearing. This has disrupted the micro-water cycle in the region.

It is apparent both on a local and global level that disruption of large areas of vegetation are causing significant and, now, unpredictable changes in rainfall distribution and duration.

There are large parts of the globe that are being deforested in order to produce more food. It is now obvious that these terrestrial disruptions are causing significant changes to the water cycle both on a local and global level. These water cycle disruptions confound the overall effects of climate change and exacerbate the problems. In order to feel the world's population future, we need to find ways to maintain significant forest cover, at the same time improving efficiencies and productions on the lands that are under agricultural management.

4.2.2   Water Use Efficiency in Irrigated Agroecosystem

Agroecosystem water use efficiency (WUE) is defined as the ratio of crop carbon gain to actual water consumption. As the pressures on water use have become greater with increased water scarcity and water quality, irrigation for agricultural production has become more expensive both economically and environmentally. Plant breeders are working hard to improve the genetic basis for crop WUE but crop management is also key to optimizing this very important aspect of production. Water use efficiency on an agroecosystem scale is one of the indicators of agroecosystem sustainability.

Irrigation methods in North China have become a very hot issue due to the use of ancient and inefficient irrigation techniques. These techniques have not only resulted in low WUE but also been responsible for increases in drinking water deficit of a nearby city (Hubacek and Sun, 2005). The reason is rooted in low-efficiency methods used by small-scale farmers with no opportunity to access new technology. In response, recently, small farmers in North China have built associations of water users; this has meant that groups of farmers have been able to access sustainable WUE-improving management technologies and techniques. This communal approach has resulted in significant progress (Hu et al., 2014; Wang et al., 2010) and is a model for other communities.

4.2.3   Social Aspects of Water from Agroecosystems

It is not enough to interpret water in our food-dependent agroecosystem just in geobio-chemical cycles within the discipline of ecology; water is a key component of the socioeconomic interactions of our society. As indicated in Fig. 4.5, human beings rely on the acquisition of renewable resources (water) and ecosystem services on Earth. Activities that improve welfare are driven by social elements and are affected by the system. However, waste and human interfering affects the ecosystem functions generated in these activities. Human social systems act as subsystems of agroecosystems. Human activities are driven by demand for ecosystem services, including water. Society anticipates a sustainable supply of such services as water availability and purity because water in agroecosystems is both a basic human need and has a role in producing healthy food for sustaining human life. While agricultural water services are just one of the many crop production inputs, it is a critical input without which intensification and diversification of agricultural production would be impossible. In addition, water misuse elicits many negative effects on the natural cycle of water in agroecosystems thus threatening sustainability. In turn, this results in social deterioration such as hunger and poverty, disease and even political instability that exacerbates the natural degradation in water (Fig. 4.5). People are dependent upon water yet they tend to degrade that very resource upon which society is built.

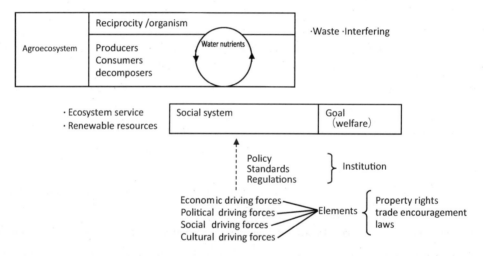

Fig. 4.5　Agroecosystem reliance on water to maintain its sustainability

Therefore, beyond the mere biological consideration of water, we need a framework of political policy informed by ecology to govern the use of water in building a sustainable food system.

4.3　Aspects of Water Management in Agroecosystems

Ecosystems naturally function as water purifiers; however, when precipitation falls on agricultural lands with altered hydrology, the water cleansing process is compromised. The intensified processes of fertilization, pesticide application, irrigation, and animal production only further stress the ecosystem and prevent it from working properly. Better management of agroecosystems to protect water supplies would include understanding quality control of soil, crop nutrients, pesticides, animal manures, and other residues, and incorporating principles of hydrology, plant cover, and stress maintenance. In the next sections, we propose a framework with five aspects for managing water in the agroecosystem.

4.3.1   Improving On-Farm Water Management

Managing root zone water application and procuring high productivity is contingent upon several factors such as soil fertility, cultivar selection, cropping density, disease and pest management, and post-harvest controls. With water restrictions on the postharvest period of crop yield, efficient water management practices become critical. With technologies that schedule irrigation, biotechnology and geographic information systems, agricultural improvements can maximize the use of a limited water supply.

4.3.2   Improving the Performance of Irrigation System

Farmers need a reliable supply of water from irrigated systems to prepare crops in a timely manner and in the right volume. While these irrigation systems were initially designed to alleviate problems of food scarcity, poverty, and unemployment, now they benefit the farmers and communities alike. Irrigation management fulfills multiple objectives, from recharging local aquifers to maintaining shelterbelts and orchards to mitigating environmental externalities associated with waterlogging and salinity. Precision application of water on a timely basis is the key issue.

4.3.3   Using Non-conventional Waters

Increasing water supply through the reuse of drainage water has become a good option for arid, semi-arid areas, and water-scarce countries. The FAO (2007) has provided guidelines covering many aspects of water conservation at the field level, water reuse at the scheme level, and safe disposal and treatment of drainage effluent. Other non-conventional uses of water to increase supply include treating wastewater, saline water, and greywater to increase water use efficiency, reduce water losses and pollution, increase recycling and access to high-quality water for water-scarce countries.

Water harvesting refers to the process of capturing, collecting, and concentrating runoff and rainwater for an available resource. Water harvesting can ease pressure on existing available resources to supply crops with rain-fed irrigation water.

4.3.4　Recycling the Water

Using recycled and rain-fed water in agriculture will allow for the allotment of more raw water for other higher utility uses. These uses include municipal supplies, environmental reserves, and hydropower generation (FAO, 2007). Addressing global water scarcity in the long run requires the emergence of new policies. These policies will rely on the pursuit of food security, investment in water-related activities, and relieving agricultural reliance on irrigation. Specifically, government expenditures will have to focus on irrigation, flood control, dams, and affiliated interventions.

4.3.5　Exploring Agroecology to Leverage Agroecosystem Water Governance

Global agriculture is facing growing challenges at the crossroad of food provision and other ecosystem services. Supply and water management in agroecosystems is in a nexus point for their tradeoffs, because at the same time needing precious water in ecosystems for food production, we sacrifice its other ecosystem services that also contribute to human welfare. Agroecology is a multidisciplinary and multiscale approach to the design and management of agricultural systems through scientific research, practice, and collective action. Under the umbrella of agroecology, we could explore Ecosystem Service as a water managerial tool to balance its provision and other services. In doing so, creating the mechanism for payment to providers of the ecosystem service of water is considered to be the first and imperative step (Ricart et al., 2019; Yousefi et al., 2017; Wang et al., 2016a, b).

Chapter 5
Linking Agroecosystems to Food Systems

C. D. Caldwell and S. Wang

> *"Eventually we will realize that if we destroy the ecosystem we*
> *destroy ourselves."*
> — *Jonas Salk*

Abstract It is essential for sustainability, no matter what system we have for food production, to have balance. Sustainability of farming systems has traditionally been one where economics and the environment are balanced; for long-term sustainability of agroecosystems, in which humans are central to the success, it is essential that we add a third concept to the balance, human health. The concept of balance can help us in our deliberations around analyzing a production input, a crop, a cropping system, a farming system, or even a value chain. The question we need to ask is whether whatever is being done produces a good balance of economics, environment, and human health. Agriculture and food are at the center of human health, economic, and environmental sustainability.

Learning Objectives

After reading this chapter and considering the topic, students should be able to:

1. List the parts of a typical food system and explain the interactions of the parts with each other locally and globally.
2. Explain how the concept of "making money and respecting the environment" is central to agroecology.
3. Explain the concept of value chain and the role of consumers in the value chain.
4. Explain the concept of "finding the sweet spot" among environment–economics–health in agriculture.

5.1 Food Systems in Association with Agroecosystem

5.1.1 Parts of a Food System (and Their Needs)

Consider the parts of a food system. It can be simply stated as the connected system of the consumer–retailer–processor/wholesaler–producer/farmer. The overall food system is a "pull" type of system; what this means is that the consumer has the power to oversee this system. If the consumer chooses to buy something, the rest of the food system needs to supply it. Therefore, the first question to ask: what does a consumer want in the way of food? Increasingly, the consumer wants to have high quality, lots of choice in products, and low price. What does the retailer want? The retailer wants to make money consistently and over time. This means that the retailer must supply the consumer consistently with high-quality products, lots of choice at a low price. In order to do that, the retailer must obtain such goods at a lower price than what they sell them for. The processor/wholesaler wants consistent profit and therefore must supply diverse, high-quality products at an even lower price. This chain brings the ultimate responsibility for high quality, diversity, and even lower price to the producer/farmer. What does the farmer want? They want to make money consistently to support their business as well. The one big difference in this model is that the "factory" for the farmer (i.e., the farm) is also the home for the farmer and family. It may, in fact, be a multigenerational farm passed down for generations. If the processor or retailer goes out of business, they lose money and a building; if a farmer goes out of business, they lose their home and perhaps a legacy. The economics of food production differs from that of widget production in this important social aspect. There are unique, significant social issues in consideration of food systems mainly nested in rural areas of countries.

5.1.2   *Problem of Disconnection*

Increasingly, consumers have little or no contact with farmers who produce the food they consume daily. Historically, when many people lived in rural areas, consumers and farmers knew each other. This communication provided trust between the two groups and feedback to the farmer on what is desired by consumers. Presently, that line of communication has been broken. Many consumers are suspicious of farmers and do not trust the methods used to grow and market their food. There is also a geographic disconnection in the food system. The average distance traveled by food from farmgate to the consumer has been variously estimated to be between 2500 km and 5000 km. This means that the energy and greenhouse gas emission for transportation is added to the already considerable carbon footprint of food production. As a result of this communication and geographic disconnection, many people have responded with buying local and supporting community-based systems. While this effort is laudable, if it is based on mistrust and misunderstandings, this needs to be addressed. Many of those who are making these alternative choices do so with an increased price and time commitment that not every family can afford. It is incumbent upon agriculture to provide safe food and a trustworthy food system to everyone and not just those who can afford it.

5.2   Agroecological Transition: Making Money and Respecting the Environment

The early farmer selected and cultivated wild plants that had desirable characteristics such as large fruit size, sweet taste, fast growth, and disease and insect resistance by natural selection. The earliest domesticated plants appear to have been cereals and legumes (such as peas, wheat, and barley); after that, the Egyptians (circa 3500 BC) developed technologies for drainage, irrigation, land preparation, and food storage. The ancient Greeks listed common plants and other medicinal usages in herbals and are credited with establishing the scientific field of botany.

The agricultural pattern has changed dramatically since the Industrial Revolution. This pattern produced an undesirable change in the physical, chemical, or biological characteristics of air, water, soil, or food that can adversely affect human health, survival, or activities and our co-inhabitants of the ecosystems.

In an ecological view, post-industrial agriculture has a high-throughput economy (Fig. 5.1); such economy requires inputs of high-quality energy and matter from nature. These resources flow through the economy and are converted to products, but also produce low quality heat energy, waste, and pollutants as by-products.

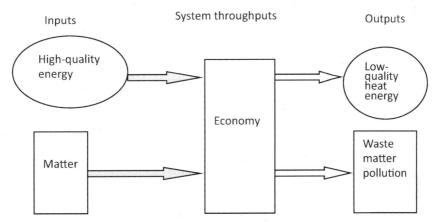

Fig. 5.1   Modern industrial agriculture is a high-throughput economy subsidized by nature and society

Value Chains

In order to address the problems of disconnection that have evolved with our food system, it is important to define the process by which food makes it from soil to shelf. We need to retrace the movement of food, who is involved and how we may increase communication and trust and decrease environmental impact. The sequence of movement of food and the addition of processing, packaging, moving, and marketing is called a value chain, since at each step in the process, value must be added in order for money to be made. So, the food value chain is really a network and not a chain. It is a network of stakeholders, participants in the food production system, who are involved in growing, processing, and selling the food that we eat. Let us again consider the participants (stakeholders in the chain):

1. Producers develop, grow, and trade food commodities such as wheat.

2. Processors, who add value to primary products by a manufacturing or marketing process to produce items such as bread or flour.

3. Wholesalers and retailers that distribute, market, and sell food.

4. Consumers, purchasers at the end of this chain.

5. Regulators that regulate and monitor value chain from producer to consumer; these may be governmental regulators or may be industry organizations for selfregulation for quality and safety.

It is important to make a distinction between a supply chain and a value chain. A supply chain is a passive description of flow of material, whereas a value chain is a more active economic approach to understanding that sequence of events, including understanding the role and impact of research and development, design, and approaches to marketing. Fig. 5.2 is an example of a value chain of shellfish from Canada to China.

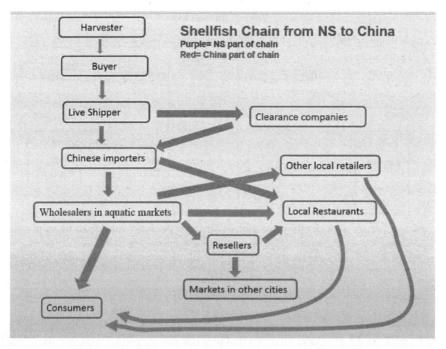

Fig. 5.2　Example of a shellfish value chain from Canada to China (Somogyi et al., 2019)

The challenge to the members of a value chain is to make it as seamless, fair, and efficient as possible. This is where **value chain management** (VCM) can become a powerful, positive tool. The goal of VCM is to organize and analyze all activities in the value chain. This involves establishing channels of communication among the stakeholders to ensure that the chain or web is one that is efficient and sustainable. With poor communication among the stakeholders in the value chain, each participant tends to take a competitive approach to the value chain and not a cooperative one. This breeds suspicion and worries about fairness; each participant in the value chain is wanting a fair return for their investment and effort. Therefore, fairness in the chain is one indicator of sustainability. The result of good VCM is a satisfied customer with cooperating stakeholders invested in the continued success of the chain. Collaboration among the various stakeholders along the food value chain is more important than ever for reasons of food safety, efficiency especially energy, reducing losses of food, waste management, and food price control.

5.3   Concept of Sweet Spot

The concept of a "sweet spot" is a simple visual tool that can help us in our deliberations around analyzing a production input, a crop, a cropping system, a farming system, or even a value chain. The key question is whether whatever is being done produces a good balance of economics, environment, and human health. Agriculture and food are at the center of human health. It is essential that, whatever system we have for food production, it is very important to have balance in order to be truly sustainable. Sustainability of farming systems has traditionally been one where economics and the environment are balanced. For long-term sustainability of agroecosystems, in which humans are central to the success, it is essential that we add a third corner, health, to the balance; hence, the illustration in Fig. 5.3.

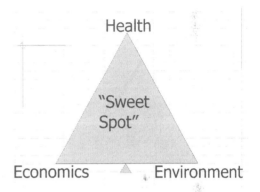

Fig. 5.3   Illustration of the concept of the sweet spot in evaluating systems

5.4   Educating Consumers with Agroecology

As described above, modern industrial agriculture is a commercial-based production system, within which agricultural products (particularly food) are produced, distributed, and consumed at discrete stages. Supermarkets are seen to be the culmination of the system; a supermarket usually indicates or hides information on agroecosystems, including what is produced in certain agroecosystems and where and how production occurs, by means of product distribution, labeling, packaging, and pricing. However, the supermarket often obscures any integration of the agricultural system. The supermarket is the great leveler: the consumer does not see the link between themselves and the complex, international industrial agriculture chain that provides the materials for their grocery basket. In particular, the consumer is isolated from the producer (farmer). This lack of communication between consumer and producer tends to breed lack of understanding of each other and eventually lack of trust (Fig. 5.4).

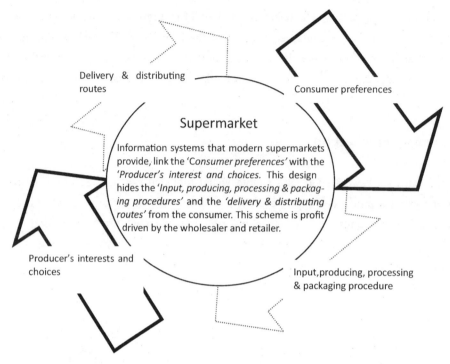

Fig. 5.4   The central role of the modern food supermarket in disintegration of the interrelated parts of agroecosystems

Many problems of modern agriculture arise from the disconnect between the consumer and the producer, in addition to the physical disconnect of the product from the agroecosystem in which it originated, by way of long-distance transport. Since production is determined by consumption (supply and demand), we must pay attention to the information that is directly or indirectly disseminated (or not made available) to consumers in supermarkets.

Modern agriculture entails increased distances between producers and consumers, planners and beneficiaries, researchers, and practitioners. People throughout the world have differing environmental views, often based on lack of information or inaccurate information. Many people in industrial consumer societies are developing a worldview of management. Advocates for environmentally sustainable economic development generally have an environmentally sustainable worldview.

A precondition to reconciliation is finding balance in views from multiple shareholders. Therefore, diverse teams are needed to deal with agroecosystems, building in knowledge and communication from both ends of the food system—in the supermarkets and from the farmers (Chaparro-Africano, 2019, Montoya and Valencia, 2018). In between, the consumers are the powerful end, because what we produce is determined by what we consume. In this regard, agroecology plays a new role in educating the consumers. In 2009, Wezel et al. redefined agroecology both as a science, as a practice and as a social movement (Wezel et al., 2009). This serves as a call to action for the network of university professors, farmers, and policymakers to steer agroecological transition worldwide (Wezel et al., 2018).

Chapter 6
Agroecosystem Health and Services

S. Wang and S. Smukler

> *"Economic deficits may dominate our headlines, but ecological deficits will dominate our future."*
> —Lester Brown

Abstract Healthy ecosystems and agroecosystems are our ultimate goal for the survival both of humanity and all creatures on earth because both natural ecosystem and agroecosystems provide goods and services without which, society would cease to exist. These goods and services range from essential activities such as keeping the environment suitable for life to offering aesthetic pleasure from the simple existence of nature. There are nine key ecosystem services discussed in this chapter: ①production of ecosystem goods, ②generation and maintenance of biodiversity, ③partial stabilization and regulation of climate, ④mitigation of droughts and floods, ⑤pollination of crops and natural vegetation, ⑥dispersal of seeds, ⑦natural pest control, ⑧services supplied by soil and ⑨provision of aesthetic beauty and intellectual stimulation that lift the human spirit. Methods of valuing and measuring ecosystem health rely on vigour, organization and resilience. Agroecosystem health branches out from ecosystem services and health, which makes it slightly more difficult to measure. In this chapter, a theoretical basis for agroecosystem management is discussed, with a model for overall agroecosystem health, microscopic management and macroscopic measurement. There is a need for a sound evaluation and potentially a payment policy to stimulate realistic, sustainable managerial function of ecosystems and agroecosystem services and health for the benefit globally.

Learning Objectives
After considering this topic, students should be able to:

1. Define ecosystem health and how it relates to ecosystem services.
2. Explain the basic indicator concepts of ecosystem health: vigour, organization and resilience.
3. List goods and services agroecosystems provide for humans.
4. Explain the difference between ecosystem services and agroecosystem services.
5. Describe the function of agroecosystem services in agricultural production.
6. Explain the differences between micro and macro management in terms of agroecosystems.

6.1　What Is Ecosystem Health and How Does It Relate to Ecosystem Services?

6.1.1　Ecosystem Health and Its Measurement

In 1990, the Environmental Protection Agency of the United States broadened its management goals from protecting human health to protecting ecosystem health (Costanza, 1992). Since then, the science advisory board has improved management to reflect linkages between human health and ecological health in the U.S. national environmental policy. Public awareness about sustainability has broadened the scope of ecosystem health into a social objective. Ecosystem health, as analogous to human health, can take on many different definitions, including but not limited to:

- Health as homeostasis
- Health as the absence of disease
- Health as diversity or complexity
- Health as stability or resilience
- Health as vigour or scope for growth
- Health as a balance between system components

In 1992, Costanza developed the following two definitions of ecosystem health: ①An ecological system is healthy and free from 'distress syndrome' if it is stable and sustainable—that is, if it is active and maintains its organization and autonomy over time and is resilient to stress; and ②the concept of ecosystem health is a comprehensive, multi-scale, dynamic, hierarchical measure of system resilience, organization and vigour. A healthy ecosystem must maintain structure (organization) and function (vigour) over time, in the face of external stress (resilience). The definition of a healthy ecosystem must also stay within the parameters of the larger system of which it is part (context) and the smaller system it creates (components).

Fig. 6.1 demonstrates an approach to measuring ecosystem health. Since measures of health are inherently less precise and difficult to obtain, this model shows the progression from 'indicators' directly measured from the status of a component, to composites of these indicators, referred to as 'endpoints', to 'values' of health.

Each component of an ecosystem, whether it is a cell, organism, species or biosphere, has a finite lifespan. A healthy and sustainable ecosystem in this context indicates a component attains its full and expected lifespan.

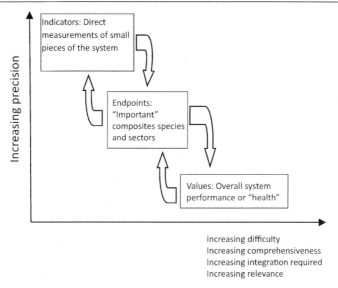

Fig. 6.1   Relationship among indicators, endpoints and values of ecosystem health (Costanza, 1992)

Cutting short the life span of an ecosystem component is an indicator of poor health. The key, however, is differentiating between changes due to normal lifespan limits and succession, and changes that cut short the life span of the ecosystem component. Distress syndrome refers to the breakdown of an entire system, in an irreversible process where the normal lifespan is cut short (Rapport, 1989). A fairly comprehensive assessment of overall system performance and health combines three basic concepts: vigour, organization and resilience. Here, Vigour is generally quantified as a measure of activity, metabolism or primary productivity. For instance, methods that already exist in the system such as gross primary productivity and gross domestic product measure the overall activity, metabolism or economic growth in the system. While these measures help evaluate vigour easily, vigour alone does not indicate overall health. Organization refers to the interconnections within ecosystems, affected by the diversity of species within the system and the exchange pathways between them (Rapport et al., 1998). Unlike vigour, organization measurements are not straightforward, they involve complex analyses such as an Input-output (I/O) analysis or an ecological/economic mass-balance model. Resilience pertains to the ability of the system to withstand stress disturbances and perturbations. The concept of resilience consists of two main aspects: ①time for the system takes to recover from stress and ②thresholds for absorbing stress in which the system can no longer recover (Zhu et al., 2012; Costanza, 2012). Measuring resilience requires dynamic simulation models for prediction, which makes true results difficult to obtain. The most accepted method of resilience measurement is the Recovery Time (R_t), or the maximum magnitude of stress (MS) divided by the Recovery Time, as an estimate of the time it takes for the overall system to recover.

A healthy ecosystem exhibits a balance among vigour, organization and resilience, whereas a system that exhibits less of one or two of these components shows signs of crystallization, brittle or eutrophic features. For instance, nutrient-rich lakes showing early succession have little organization and should be considered 'eutrophic'. Ultimately imbalanced, unhealthy ecosystems hinder the human economy, as remediation requires trillions of dollars and often still cannot offer the full range of ecosystem services naturally derived.

Just like 'Sustainability', without the practical measurement, ecosystem health is just an abstract slogan, essentially, a healthy ecosystem should sustainably provide a range of ecosystem services (Costanza, 2012).

6.1.2 Ecosystem Services and Their Measurements

Ecosystem services (EEs) is now a well-defined and active field of ecological economics since it was proposed by Ehrlich and Mooney in 1983, and after first quantitively measured by Costanza et al. in 1997 in particular. Basically, EEs are comprised of both goods and services that ecosystems provide to our humankind (Tomich et al., 2011), and have been classified into four categories: provision, supporting, regulating and cultural services (MEA, 2005).

In sum, EEs sustain all species (including humans) by providing the fundamental life-supporting services such as oxygen production, primary plant production, water and all habitats. EEs can also be considered as natural capital allied to the ecosystem, in contrast with social capital which is allied to the social system (Daily et al., 1997), and then functioning of the world's economies. However, we need to know that our human economies are merely a subsidy of an ecosystem; a sustainable society lives off the biological income provided by EEs without depleting or degrading them. Theoretically, all ecosystems could operate sustainably by using renewable energy and recycling chemical nutrients, Fig. 6.2 shows a conceptual diagram of this process, in such, EEs provide the dialogue platform for ecology and economics.

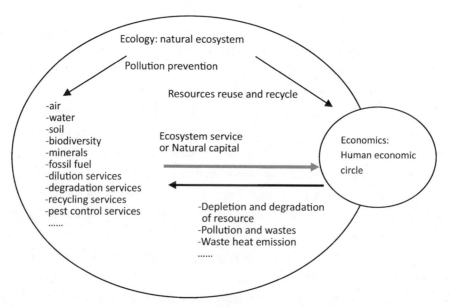

Fig. 6.2　Ecosystems services (or natural capital) sustain all species and human economies

As such, Ecosystem services can also be classified into four categories: provisioning, supporting, regulating and cultural services, briefly listed as follows (Wratten et al., 2013):

Provisioning services, including food and other materials for human consumption, usually being valued by and purchased from the marketing constructed with neo-classical economics. These marketable services are categorized into seven groups, namely: foodstuff, site views, power, recreation, materials, fuel, others (Fig. 6.3, Wang, 2007).

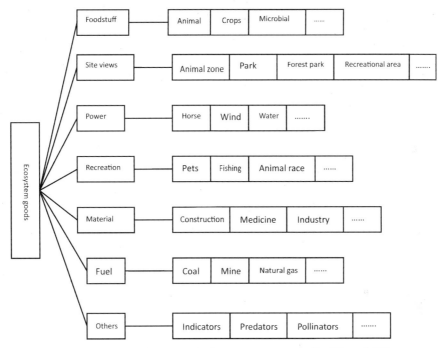

Fig. 6.3   Production of goods that ecosystems provide (Wang, 2007)

- Supporting services, ecosystem functions that support the production process other than provision services, for instance, beneficial insects which play a great role in balancing populations in field level, and bees that help the pollination and nitrogen-fixing plant that facilitate nutrient cycling, and now often are substituted with external inputs (pesticide, fertilizer and mechanical power) in conventional agriculture systems.

- Regulating services, ecosystem function that regulate the essential ecological processes and life-support systems through their biogeochemical cycles and other biospheric processes, for example, hydrological flow in the plant-soil-atmosphere plays a critical role in arable farming by regulating the climate and water condition.

- Cultural services, ecosystems also provide recreation, aesthetics and education that beneficiary to human health and well-being. An important but often misunderstood ecosystem service is the aesthetic beauty, intellectual stimulation and renewal of spirituality humans receive from nature. These types of ecosystem services provide people with the opportunity to partake in activities such as hiking, camping and gardening, to observe and internalize nature through art, film and bird watching. A long history of religion, art and cultural traditions highlight the deep appreciation people have for the peace and beauty one may find in nature.

EEs presented above highlight the many benefits that the natural ecosystem provides to humanity. EEs support life and maintain the human economy. These free services would otherwise cost humanity billions upon billions of dollars, in addition to the benefits they provide through health and a good quality of life. Hence, to protect and maintain the EEs, the first step is to value them.

Valuing ecosystem services proves extremely complex and highly uncertain. Generally, measurements rely on marginal values, such as estimating an ecosystem flow or service in terms of how much preservation or destruction would occur due to the loss of a specific area. This qualitative valuation is enough evidence to validate or reject plans on land altering projects. However, replicating or replenishing lost ecosystem services generally far exceeds the estimated worth of those services (Figs. 6.4 and 6.5).

The importance of valuating ecosystem services is threefold: ①to aid in macroallocation decisions between economic production and ecosystem services and between economic infrastructure (including agriculture) and ecological infrastructure; ②to avoid exceeding economic thresholds, where ecological degradation costs will exceed the benefits; and ③to avoid exceeding ecological thresholds that threaten long-term sustainability and intergenerational macro-allocation of services.

The value of ecosystem function, the capacity of natural processes and components to provide goods and services that satisfy human needs, depends greatly on the value of the ecosystem services.

The value of an ecosystem service stems from its ecological, socio-cultural and economic importance. These services can be measured by its use value and non-use value. Fig. 6.4 depicts a framework for mapping and valuing ecosystem services based on two key concepts: ①valuation—how can we assess the relative value of various ecosystem services? and ②incentives—how can we provide rewards for providing ecosystem services? Fig. 6.5 explains the valuation process further in terms of use and non-use values.

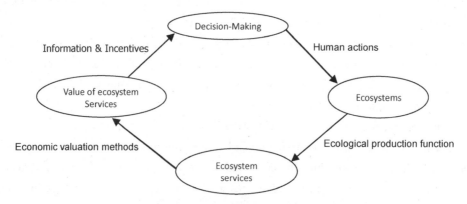

Fig. 6.4　A framework for mapping and valuing ecosystem services

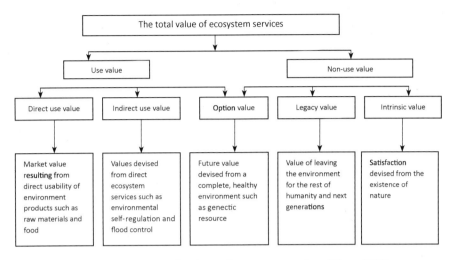

Fig. 6.5   Value classifications of ecosystem services (Wang, 2007)

Use value in Fig. 6.5 refers to direct use value and indirect use value. Direct use values include market and trade values, usually for goods, but can also apply to information and regulation. Indirect use values, on the other hand, impart valuation techniques through indirect assessments. Usually, indirect use value measures the value of the loss of a service. These measurements include avoided costs, replacement costs, factor incomes, travel costs and hedonic pricing (de Groot et al., 2002). For instance, flood control is considered an avoided cost or the value people are willing to pay to avoid damages from floods.

6.1.3   Integrating Ecosystem Service with Ecosystem Health Through Ecosystem Management Frameworks

Ecosystem health and ecosystem services are connected; comparatively, EEs are easier to link to policy-making, research and education processes then is ecosystem health. However, is actually more feasible to main ecosystem health, in order that ecosystem services can be maintained (Ford et al., 2015). Therefore, in general, it is imperative to add the complex ecosystem by combining ecosystem services with ecosystem health as an integral system (Kang et al., 2018).

From the managerial view of human society, ecosystem management should offer a bridge to integrate the goals of ecosystem health and ecosystem services go to sleep wake up together (Costanza, 2012).

Ecosystem Management integrates scientific knowledge of ecological relationships within a complex socio-political and values framework towards the general goal of protecting native ecosystem integrity over the long term (Grumbine, 1994); Pavlikakis and Tsihrintzis (2000) developed the following steps to EM methodology (Fig. 6.6):

- Localization of issues: determining the most important issues.
- Participation of the population: public participation in the decision-making process.

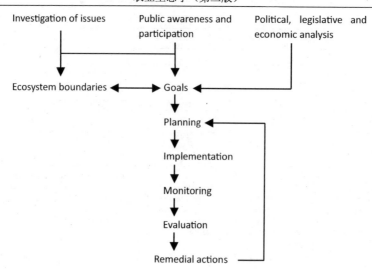

Fig. 6.6   General steps in ecosystem management methodology (Pavlikakis and Tsihrintzis, 2000)

- Political, legislative and economic analysis: conducting political, legislative and economic analysis before beginning process.
- Definition of goals: stating clear goals and communicating them to the public.
- Definition of the boundaries of the ecosystem: understanding local restrictions and opportunities; understanding the needs and expectations of local residents.
- Development of a plan: involving all entities in the planning and communication process; securing funding and human resources.
- Monitoring: collecting high-quality scientific data and information.
- Evaluation: continually evaluating the EM project to reach the stated and expected goals.

Under this framework, considering soil is critical to both ecosystem health and service maintenance, a Soil stewardship strategy as an example could be developed in Fig. 6.7.

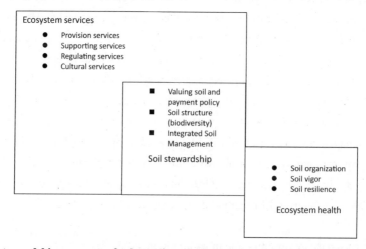

Fig. 6.7   Soil stewardship as a nexus for better integration between Ecosystem Services and One Health

(Keith et al., 2016)

6.2   Evaluating Agroecosystem Health and Services

6.2.1   Agroecosystem Health

Agroecosystems are indispensable components of the global ecosystem. Unsurprisingly then, agroecosystem health plays an important role in agricultural stability and sustainable development of human society. Human participation differentiates agroecosystems from all other ecosystems, allowing for a complex integrated nature–economy–society system that naturally changes over time and space. Ideal agroecosystem health accounts for variation with time and space without limiting the system's ability to uphold structural and functional characteristics to produce agroecosystem goods and services (Zhu, 2011).

Research on agroecosystem health provides a scientific basis and diagnostic tools for monitoring and managing agroecosystems. Rational human intervention can enhance healthy system dynamics and sustainable development, while ill-informed human intervention may interfere with overall system health, resulting in damage, ecosystem degradation and ecological disaster. An agroecosystem management model must combine sound ecology, biology, economics, information science and other disciplines. Agroecosystem health and agroecosystem management highly depend upon each other and require a symbiotic relationship to pursue sustainable agricultural development.

Broadening the ecosystem health assessment method based on the dynamic characteristics vigour, organization and resilience (maintenance), in agroecosystem health we also assess structural, functional or organizational characteristics.

Structural components include resource availability, accessibility, diversity, equitability and equity. Functional criteria include productivity, efficiency and effectiveness. Organizational criteria include integrity, self-organization, autonomy and self-reliance. Generally, measurements of agroecosystem health recognize important linkages between agroecosystem health and soil and water quality; human health; Integrated Soil Management (ISM) and Integrated Pest Management (IPM); ecological impacts of biological indicators, genetically modified crops (transgenic crops); roles and impacts of agricultural inputs, policy and landscape ecology; and green food development (Peterson et al., 2017).

6.2.2   Agroecosystem Services: Classification and Evaluation

As we recall from Chaps. 1 and 2, agroecosystems are critically important managed ecosystems. Ecosystem services from agricultural land provide important non-market goods and services imperative to sustain life at all levels and agriculture. Therefore, maintaining and restoring ecological functions to an agroecosystem is of paramount importance for agricultural sustainability.

Agroecosystem services build upon ecosystem services discussed earlier in this chapter, as such, agroecosystem service consists of provisioning, regulating, supporting and cultural services, but all having specific utilizations. Some of the key ecosystems goods and services in agroecosystem are exemplified as follows:

6.2.2.1　Production of Ecosystem Goods

Agroecosystem ecosystems provide people with daily goods and services that are crucial to human survival. Goods of this sort include foods, marketable items, labour and materials. Ecosystems offer food in the form of vegetation, fruit, nuts, spices and fish for human consumption. Fish plays a major role in the economy through commercial harvest, sale, sport fishing and other fishing related activities.

Agroecosystems also offer marketable goods that stem from animals using grasslands as habitat. These goods include wheat, oats and barley, animal meat such as deer, moose and elk, and animal products such as wool, leather and milk. Animal labour can further be considered as a secondary product of ecosystem services.

Materials derived from agroecosystems, such as fibre, fuelwood and industrial products also play a major role in society. Most people have become heavily dependent upon these natural materials offered by ecosystems

6.2.2.2　Generation and Maintenance of Biodiversity

The generation and maintenance of biodiversity is a crucial component of ecosystem services (further information available in Chap. 10). Generating and maintaining a multi-scale, interdependent, co-evolutionary array of organisms ensures humans continually benefit from the abundance of biological diversity. Biodiversity supports conventional crops and future food security, genetic and biochemical resources and pharmaceutical enterprises, which exceed $40 billion per year for nearly 80% of the medical systems worldwide (Daily et al., 1997).

6.2.2.3　Partial Stabilization and Regulation of Climate

Ecosystems help stabilize climate on the global scale and regulate weather and temperature on the regional scale. Natural ecosystems stabilize climate through the prevention of overheating and by the removal of excess greenhouse gases in the atmosphere. The stabilizing mechanisms usually result from natural trends in feedback mechanisms. On a regional scale, ecosystems exert influence over temperature and weather, influencing precipitation, evaporation and transpiration.

6.2.2.4 Mitigation of Droughts and Floods

Soil, plants and plant litter soak up rainfall each year, allowing water to permeate through the soil and plant roots into aquifers and streams (Daily et al., 1997). Protection from soil and vegetation prevents annual rainfall from flooding most of the Earth's surface. On the other extreme, when rainfall is scarce, soil and plants store water as groundwater and in plant roots to nourish the ecosystems, reducing disruptions in the water cycle and loss of nutrients.

6.2.2.5 Pollination of Crops and Natural Vegetation

Ecosystems provide habitat for natural pollinators (usually animals) for plants and crops. Pollinators, from bees and hummingbirds to beetles and birds, require many different habitats to complete the different stages of their life cycles. Keeping these ecosystems intact ensures over 200,000 species of plants receive pollination. Likewise, many agricultural crops require pollination from wild animals as well. Natural ecosystems provide this service for free; however, if humans were to try and replicate this process it would cost many billions of dollars each year.

6.2.2.6 Dispersal of Seeds

To disperse seeds for germination in areas beyond a plant's rooted zone, plants depend on other mechanisms in an ecosystem such as wind, water and animals. The wind carries seeds such as dandelions, while water moves seeds such as the seafaring coconut. Animals, on the other hand, transfer seeds in a variety of ways. Some seeds have evolved to latch onto animals and get carried away; other seeds are disguised as sweet fruits that pass through an animal's digestive tract. For instance, the Southern Cassowary in Australia disperses over 200 seeds species many kilometres away from the original plant roots. Other seeds have evolved to have a very specific counterpart, such that only one animal species can disperse the seeds. This ecosystem service provides numerous ecosystem benefits, particularly by protecting plant species from extinction due to natural or human disturbances through multiple seed dispersal mechanisms.

6.2.2.7 Natural Pest Control

Pests, competitors of food, fibre and materials, destroy much of the world's harvest and compete for water, light and nutrients. Luckily, natural pest controls in the form of birds, bugs, diseases and other organisms act as biological control agents to limit pest abundance. No chemical pesticide truly competes with a biological control agent, since pests can quickly evolve to resist synthetic poisons acting as artificial prey.

6.2.2.8　Services Supplied by Soil

Soil provides four interdependent natural ecosystem services (also see Chap. 3):

- Soil shelters plant seeds for germination and provides physical support for plants to anchor their roots and grow.
- Soils maintains the cycling and movement of nutrients to plants acting as a buffer to fertilizer and leaching and giving plants access to a supply of nutrients when needed.
- Soil plays an important role in the detoxification and decomposition of dead organic matter and wastes, breaking down chemical bonds through specialized reactions.
- Soil regulates carbon, nitrogen and sulphur stored in vegetation.

6.2.2.9　Provision of Aesthetic Beauty and Intellectual Stimulation That Lift the Human Spirit

At the field level, provisioning services in agroecosystems include the six "Fs"— food, fuel, fibre, fodder, fish and forage; this class also includes other tangible goods produced in agroe- cosystems and marketed in our social system such as genetic resources, biochemicals, natural medicines and fresh water. Regulating services include pollination, water regulation and purification, erosion regulation, disease regulation, pest regulation and flood control. Cultural services consist of cultural diversity, spiritual and religious values, inspiration, aesthetic values (scenic qualities), recreation and tourism. Supporting services includes soil formation, primary production, nutrient cycling and water cycling. All those services could be evaluated by current methods that have been developed for evaluating ecosystem services, but need to implement case by case (Porter et al., 2009; Sandhu et al., 2008, 2012, 2015, 2018; Makovnikova et al., 2016; Sagie and Ramon, 2015); among them, 'Willing to pay (WTP)' and 'Willing to accept' are the persuasive methods with the consideration of eco-social complexity of agriculture worldwide (Novikova et al., 2015).

6.2.3　*Agroecosystem Management by Designing Payment of Agroecosystem Service*

The management of agriculture and agroecosystems has huge effects on the quality and quantity of agroecosystem services that contribute to human production. Suitable management practices ensure agroecosystem health and the provision of agroecosystem services.

Agroecosystem management derives from a holistic perspective of agroecosystems, including all environmental and human elements. From this point of view, agroecosystem management aims to attain sustainable use and management of natural resources accomplished by using social, cultural, economic, political and ecological methods. Sustainable use and management refers to sustainable agricultural production, an ecologically based assessment of structure, function and multi-dimensionality, and spatial scale of food systems. Combining these principles, agroecosystem management can provide ecological guidelines to design and manage sustainable agroecosystems.

In theory, agroecosystem management should be highly applicable to different types of agroecosystems and easily implemented. In practice, however, the application of agroecological principles proves challenging. Maintaining a resource with minimal outside artificial inputs, good nutrient cycling within the system with few 'leaks', control of pests and diseases through internal regulating mechanisms, and resilient to human and harvest disturbances requires a balance among vigour, organization, structure, resilience and equality. One agroecosystem health assessment model established uses a four-stage framework with 36 indicators (Zhu et al., 2012):

- Level 1. Object: evaluate objectives and composite index of agroecosystem health.
- Level 2. Item: determine the subsystem of agroecosystem.
- Level 3. Evaluation factors: determine specific elements for each item.
- Level 4. Index level: detail indicators to express evaluation factors.

An agroecosystem management model should also develop in consideration of local conditions, with a design adapted to the local environment making full use of information technology, biotechnology and ecological engineering technology to carry out Integrated Plant Nutrient Management, Soil Management, Water Management and Pest Management. Conceptually, agroecosystem management must focus on desired objectives, microscopic and macroscopic management (Zhu et al., 2010).

Microscopic management in an agroecosystem refers to the monitoring and regulation of the interactions and energy flows within the ecosystem. Optimal behaviour of agroecosystems depends on the level of interactions between various biotic and abiotic components occurring in the soil, water and air. In addition, energy flows and nutrient cycles within the air, water and soil also form a basic component of agroecosystem services. In essence, microscopic agroecosystem management should effectively regulate these flows to achieve the most efficient use of energy and resources in an agroecosystem, managing both microscopic organisms to fieldwide species.

Macroscopic agroecosystem management must account for the larger ecosystem in which it exists (context). Macroscopic management practices go beyond the relationships found on the farm, to integrate into the larger economic, systemic, social and cultural driving forces, linked by policies and connecting producers and consumers. Thus, a good agroecosystem management strategy can achieve its goals through implementing decision-making, legal, participatory and ecological mechanisms.

In practice, since the current agriculture has been an indispensable part of global and national economies, the degradations of agriculture worldwide root in its disproportionate economic benefits in the global market under the current prevailing neo-economics. It is time to mainstream agroecosystem services by designing a sound payment policy of agroecosystem services into good agricultural management scheme (Henriquez and Molina-Murillo, 2017; Sandhu et al., 2016).

In summary, healthy ecosystems and agroecosystems are our ultimate goal for the survival both of humanity and all creature's on earth, because both natural ecosystem and agroecosystems provide goods and services without which society would cease to exist. These goods and services range from essential activities such as keeping the environment suitable for life to offering aesthetic pleasure from the simple existence of nature. There are nine key ecosystem services discussed in this chapter: ①production of ecosystem goods, ②generation and maintenance of biodiversity, ③partial stabilization and regulation of climate, ④mitigation of droughts and floods, ⑤pollination of crops and natural vegetation, ⑥dispersal of seeds, ⑦natural pest control, ⑧services supplied by soil and ⑨provision of aesthetic beauty and intellectual stimulation that lift the human spirit. Methods of valuing and measuring ecosystem health rely on vigour, organization and resilience. Agroecosystem health branches out from ecosystem services and health, which makes it slightly more difficult to measure. In this chapter, a theoretical basis for agroecosystem management is discussed, with a model for overall agroecosystem health, microscopic management and macroscopic measurement. There is a need for a sound evaluation and potentially a payment policy to stimulate realistic, sustainable managerial function of ecosystems and agroecosystem services and health for the benefit globally.

Part III

Digging Deeper into Agroecosystems

Chapter 7
Agroecology and Hunger

C. D. Caldwell

> *"Peace begins when the hungry are fed."*
> —Dorothy Day

Abstract In the application of agroecology, the primary goal is to benefit humans without diminishing the environment. This is especially true in terms of food production. In 2020, approximately 795 million of our fellow citizens are without enough food to live healthy, active lives. This level of hunger is principally due to five factors: lack of food production, postharvest losses, poor food distribution, poverty, and governance. Most famines throughout history can be linked to wars and conflict. These lead to the displacement of people, disruption of food production, and the inability to move food from the field to consumers. Such political and social upheavals are further complicated by any natural weather-related disasters, such as drought or floods. Agroecology addresses food value chains from soil to consumer when considering issues of malnutrition. The term food insecurity relates to more than just the presence of hunger. It involves an ongoing struggle for sufficient food characterized by considerable emotional stress and personal compromise. Food insecurity is a global concern, with the population continuing to rise along with increased rates of per capita consumption. Confounding this problem are the everincreasing effects of climate change. To meet rising food demands, we need a new "truly-green" Green Revolution. Norman Borlaug was the hero of the first Green Revolution. More heroes must emerge from the ranks of at least five specialties: plant breeders, irrigation technologists, agronomists, value chain managers, and consumers.

Learning Objectives

After studying this topic, students should be able to:

1. Define the following terms:

   (a) Hunger
   (b) Malnutrition
   (c) Food insecurity

2. Discuss the reasons for hunger in the world.
3. Describe the health and socioeconomic impacts of postharvest losses.

4. Explain the impacts of the first Green Revolution.

5. Critically discuss the options for a new #2 Green Revolution.

7.1 Hunger Concepts and Definitions

For most people living in developed nations, the term hunger describes a bit of discomfort signalling that it is time to eat again. However, there are millions of people worldwide, many of them children, living in a constant state of physical and emotional distress due to poor nutrition. This level of global hunger needs to be addressed effectively and sustainably. The primary goal of agroecosystems is to feed humans. Yet, in 2020, approximately 795 million of our fellow citizens do not have enough food to lead a healthy, active life.

7.1.1 Famines

Fig. 7.1 shows the millions of famine victims worldwide since the 1860s. Hunger in humans is caused by five factors: lack of food production, postharvest losses, poor food distribution, poverty, and/or poor governance. Malnutrition and food insecurity often result from poorly developed or executed food systems. However, data from the figure shows that famines are mostly the result of actions of society occurring outside of the food system, often through inappropriate policies and human conflict. While there were natural factors that decreased production (including drought and poor weather), the overwhelming causal factor was revised agricultural policies that failed to recognize the related impacts of their implementation. In other examples in Fig. 7.1, it is obvious that war and conflict caused the displacement of people, destroying their means of growing food and disrupting the transportation of food from production areas to where it was needed. Prevention of famine is not the main focus of this chapter. We shall not attempt to address the issues of poor governance and wars which have been the hallmark of famines throughout history.

7.1.2 Malnutrition

While the five factors mentioned above affecting hunger are interrelated, agroecology primarily addresses food value chains from soil to consumer when addressing the issue of malnutrition and other food-related challenges.

Malnutrition is defined as deficiencies, excesses, and imbalances in the caloric intake and/or nutrient quality of food consumed. Indications include disproportionate weight for age (stunted growth or underweight), low weight for height (wasting), and symptoms of a lack in micronutrient, vitamin, or essential mineral for growth and development. A second type of malnutrition becoming prevalent is due to excessive intake of calories, which can be accompanied by poor nutrient balance.

Symptoms include obesity, and diet-related disease (cardiac dysfunction, stroke, diabetes, and some cancers).

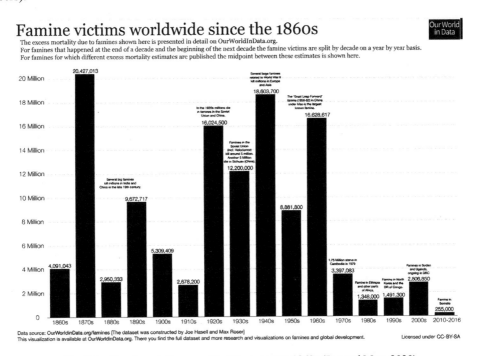

Fig. 7.1   Famine victims worldwide since the 1860s (Joe and Max, 2020)

The effects of malnutrition can be seen worldwide. Approximately 460 million adults are underweight, but almost two billion are overweight. A disturbing trend is that children are now becoming overweight or obese, with more than 40 million under the age of five being classified as such. At the same time, 160 million children are stunted, and 50 million are suffering from wasting. For both children and adults, it is not just that there is not enough food, but what is available lacks essential nutrients. For example, almost 30% of women of reproductive age around the world have anemia, a condition caused by lack of adequate iron in the diet.

In many cases, the root cause of malnutrition is poverty. People cannot access sufficient protein sources, such as legumes or meat, or the carbohydrates, vitamins, and minerals found in vegetables and fruit. At the same time, foods and beverages high in fat, sugar, and salt are both affordable and available. Obesity is rising in both rich and poor countries. It is possible to see both undernourished and overweight people in the same household, and in fact both may be present in one individual. Too many calories may be consumed yet provide too few nutrients.

As a result of global efforts, between 2005 and 2015 there was a steady decrease in malnutrition. However, since then, the numbers of malnourished and undernourished people are on the rise, reaching levels cited for 2011 (Fig. 7.2). The term "undernourished" describes those whose intake regularly provides less than the minimum energy requirements, deemed to be about 1800 kcal per day. The exact figure is determined by a person's age, height, activity level, and any physiological conditions, such as illness, infection, pregnancy, and lactation. To put this number into context, consider the McDonald's restaurant franchise, where a deluxe breakfast provides 1220 kcal, and a glass of orange juice 250 kcal. Almost a day's energy can be consumed in 20 min, with a return of hunger in a few hours.

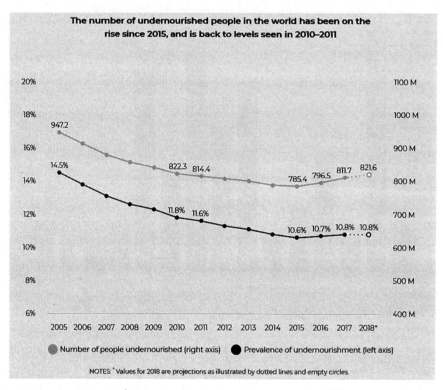

Fig. 7.2　**The number of undernourished people in the world has been on the rise since 2015** (FAO, 2019)

7.1.3　*Food Insecurity*

Food insecurity is not just a lack of food; it is more than just the presence of malnutrition and hunger. It involves an ongoing struggle accompanied by emotional stress and personal compromise. In general, three categories of food insecurity are recognized:

1. Marginal—where families worry about running out of food, and/or have limited food choices due to poverty.
2. Moderate—where people face significant uncertainty about their access to food. This means that they, at least at times, must reduce the quantity or quality of their food because of limited finances. Their access is inconsistent, giving rise to negative nutrition and health impacts.
3. Severe—where individuals reduce food intake, missing meals, and at times going days without food. This obviously puts their health, well-being, and ability to function in families, community,

and society in general, at great jeopardy.

Fig. 7.3 illustrates global moderate and severe food insecurity. It is closely linked to poverty and social inequity, but it is not experienced solely in developing countries. In fact, every country in the world has some level of food insecurity. As an example, the Vanier Institute (Vanier Institute, 2019) reports that 23% of Canadian youth under the age of 18 say they go to bed hungry at times because there is not enough food in their home. In the same year, children and youth made up 20% of Canada's population and constituted 35% of those accessing food banks. As in most other countries, food insecurity in Canada is not evenly distributed. (Tarasuk et al., 2013) In 2018, the northernmost territories, mostly inhabited by indigenous peoples, reported a level of 46% of households with food insecurity. This is a worldwide issue relating to economics, poverty, and inequality. Even those countries able to achieve improved economic growth with vibrant poverty reduction programs, may not have their efforts translate into improved food security and nutrition. Nor does improved food security necessarily translate into improved nutritional status. There remains the need to address food malnutrition globally.

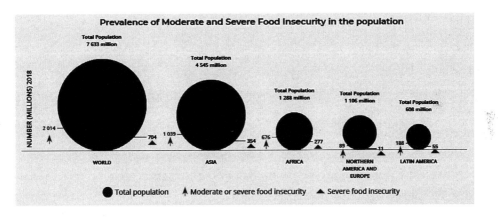

Fig. 7.3   Prevalence of moderate and severe food insecurity in different parts of the world (FAO, 2019)

7.2   Causes for Hunger

7.2.1   Food Production

In spite of global food production that is greater than ever, it remains inadequate. There are many factors limiting productivity, including the following:

– Natural factors, such as drought, complicated by effects of climate change
– Inappropriate or inefficient farming methods
– Land use conflicts, such as urban expansion and diverting food production into biofuel crops

– Overfishing

– Land and water degradation

There is a finite amount of land suitable for food production, and while its reduction seems inevitable, increasing it may be counterproductive environmentally. At one time, there were vast tracts of land dedicated to agriculture. With priorities altering, the best land in the world is being taken over by cities and covered in concrete. In other cases, agricultural lands have been flooded to provide hydroelectricity. Moving agriculture into marginal lands has caused environmental damage. It is obvious that we must be more efficient in our use of our best agricultural land in order to save much of the natural ecosystems around us, at the same time providing sufficient food. See Box 7.1 for a simple exercise to illustrate the fragile nature of the soils upon which we depend for all of our food.

> **Box 7.1　Simple Exercise: Compare the Earth and Soil Availability to an Apple**
>
> Suppose......
>
> *How much soil is there?*
>
> Think of the earth as an apple. Slice an apple into quarters and set aside three of the quarters. These three pieces represent the oceans. The fourth quarter roughly represents the Earth's total land area.
>
> Slice this "land" in half. Set aside one of the pieces. The portion set aside represents the land area that is inhospitable to people (e.g., the polar areas, deserts, wetlands, very high or rocky mountains). The piece that is left is land where people live but do not necessarily grow the food needed for life.
>
> Slice the 1/8 piece into four sections and set aside three of these. The 3/32 fraction set aside represents those areas too rocky, wet, cold, steep, infertile to actually produce food. They also contain the cities, suburban sprawl, highways, shopping centers, schools, parks, factories, parking lots, and other places where people live, but do not grow food.
>
> Carefully peel the 1/32 slice of Apple. This tiny bit of peeling represents our arable land, the land upon which we depend for food.
>
> Estimates suggest that we lose 25 billion tons of precious topsoil each year from erosion, yet we must feed an additional 71 million people each year on this diminishing resource.

The growing imbalance between supply and demand for food resources is illustrated in Fig. 7.4, showing data on cereal production utilization and stocks in storage from 2009 to 2020. The graph shows that global stocks of cereal grains in storage are either stagnant or decreasing. Present stocks are only 31% of utilization, which means that we have less than 4 months of stored cereal grains as a buffer against disastrous production in any 1 year. Demand is increasing as the population grows, and there is increased consumption due to current dietary trends. As wealth rises in any country, there is a correlated increase in demand for meat. Fig. 7.5 shows the amount of feed required to produce 1 kg of meat or milk. As societies move further up the food chain, the demand for cereal grains grows, and the overall food system becomes less efficient in terms of converting light energy from the sun into human energy. In other words, due to increased demand by increasing population, plus increased demand for meat and dairy, not to mention diversion of land into biofuel crops, the overall demand for cereal grains has recently risen by about 2.5% annually. The production of cereal grains has profited from better varieties and management practices overall. However, depending on the species and variety, the increase in production is between 1% and 1.5% annually. The gap must not be allowed to widen.

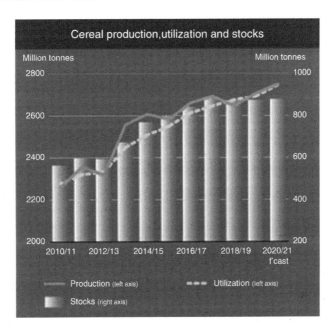

Fig. 7.4   Cereal production, utilization and stocks from 2009 to 2020 (FAO, 2020)

7.2.2 Postharvest Losses

Postharvest, the integrity of the crops must be maintained through efficient methods of storage, transport, and packaging. Losses sustained in harvested material, due to physical damage, spoilage, and destruction (such as that caused by insects and rodents) significantly disrupt the food chain. Considerable losses can be incurred at several points in the path from the field to the consumer. The Food and Agriculture Organization (FAO) has estimated that globally as much as 45% of fruits and vegetables are lost during this process. Fig. 7.6 shows the losses along the food chain, and compares the losses in both the developed and developing countries, illustrating the global nature of the problem. Fig. 7.7 shows how the losses within the commodity sectors differ regionally. This clearly indicates that there is an opportunity for positive change.

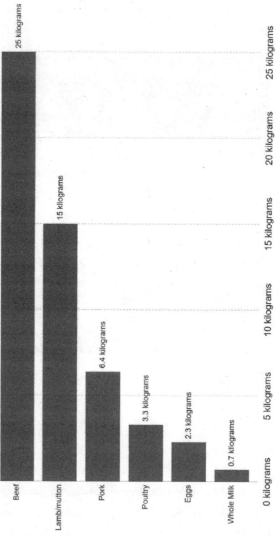

Fig. 7.5　Feed required to produce 1 kg of meat or dairy product (Joe and Max, 2020)

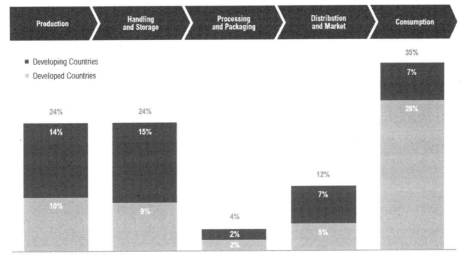

Fig. 7.6   Food losses and wastes (in % of kcal) along the stages of the value chain (Kipinski et al., 2013)

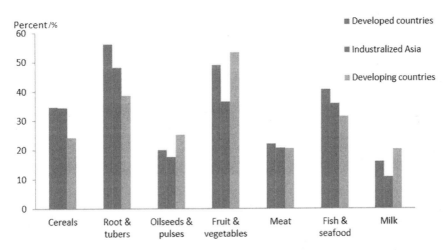

Source: Author calculations using loss parameters from Gustavsson et al. (2011) and 2009
production data from the FAO.
Note: Bars denote percent of production lost for respective commodity.

Fig. 7.7   Food losses by commodity and country (Kipinski et al., 2013)

The problem of postharvest losses is not a new one, but finding solutions becomes more urgent as pressure on food production grows and a more demanding population increases. Postharvest losses refer to reductions in both the quantity and quality of any given product. This definition takes into account deterioration which may not lead to its absolute destruction but may reduce its usage. For example, if a grain crop is affected by frost or heat, or damaged by insects, it may no longer be able to be processed into bread flour. In fact, it may no longer be suitable for human consumption, and need to be downgraded into animal feed. While salvageable, this is still an obvious loss within the food chain, both a nutrient loss and an economic loss. For a small farmer, the family loses nutrition and income.

If postharvest losses could account for losing almost half of a commodity's production, as the FAO estimated with fruits and vegetables, surely this is indicative that solving this issue is a critical part of any global food security strategy. More high-quality food leads to healthier people, and more income in the pockets of those who work all along the food chain, especially the small farmer.

7.3　Green Revolutions

7.3.1　Green Revolution #1

The period following the Second World War was a time of political and social change worldwide. At the same time, there was a great struggle for nations to provide enough food for their populations. Widespread famines were experienced. Industrialized nations had adopted new technologies aimed at increasing yields and the efficiency of agricultural production systems, especially in cereal grain crops. Conventional methods of targeted breeding aimed at improving yields and quality. Cultivation techniques that included irrigation, synthetic pesticides, and the application of inorganic fertilizers resulted in significant increases in cereal crop production.

The Green Revolution was an initiative by the industrialized nations to move these technologies into developing nations to transfer the skills, knowledge, and techniques needed to stop the hunger and starvation. Between 1950 and the late 1960s, these efforts succeeded in having a substantial impact on agricultural production. Farmers in developing nations started to adopt high-yielding varieties in association with chemical fertilizers, synthetic pesticides, irrigation, and increased mechanization. They were able to double their cereal grain production by 1985.

One great hero of this Green Revolution was Norman Borlaug, a plant breeder who was at the forefront of developing high-yielding varieties and promoting modern management techniques. For his efforts, he received the Nobel Peace Prize in 1970. This is quite an accomplishment for a plant breeder! Later in his life, Dr. Borlaug continued to devote his efforts to further improve grain production in Asia.

The overall result of these Green Revolution programs was the saving of hundreds of millions of lives from starvation. This is an end result to be celebrated in anyone's eyes. However, in the present day, the Green Revolution has been characterized as a great mistake, from both a socioeconomic and an environmental standpoint. The production packages that were transferred from industrialized nations into developing nations were dependent on high energy use, intensive irrigation, and greatly increased chemical use. With good intentions toward replacing traditional agriculture with scientific agriculture, the approach ignored human and environmental impacts that accompanied such a rapid change. Having to purchase greater quantities of inputs led to debt, with smaller farmers losing not only their business but also their land and home. Modern techniques of irrigation and chemical inputs were dependent on a larger land base to achieve efficiency, so large farms became larger, and again, small farmers were forced off their land. Resulting migration from rural to urban areas caused significant socioeconomic stress on the countries involved. From an environmental standpoint, there was often overapplication of fertilizers and pesticides with ensuing health issues and water contamination.

On the one hand, there was an increase in food production, and in many cases an increase in GDP per capita for the nations involved. On the other hand, the gap between rich and poor grew, and negative impacts on health and the environment also increased. With no social net in place, many suffered, and the "Revolution" has had an ongoing impact on those countries since that time.

7.3.2   *Green Revolution #2*

The Agricultural Institute of Canada defines "agricultural sustainability" as an option for balance, as follows:

"The application of husbandry experience and scientific knowledge of natural processes to create agriculture and agri-food systems that are economically viable and meet society's need for safe and nutritious food and vibrant rural communities, while conserving or enhancing natural resources and environment."

The first Green Revolution is credited with saving perhaps one billion people from starvation. It accomplished this by applying husbandry experience and scientific knowledge and was able to create new agriculture and agri-food systems in the countries affected. However, where it went wrong was the initiative transferred systems that were not economically viable, did not meet society's need for safe and nutritious food, did not create vibrant communities, and in the end, did neither conserve nor enhance natural resources or the environment. Therefore, using the above definition, the Green Revolution was not sustainable.

In round one, important lessons were learned. To meet current food production requirements, another Green Revolution is needed, and needed in half the time, if we are to solve the problem of a growing global population with increased rates of per capita consumption, all within the emerging confounding factor of climate change. We will need more heroes like Dr. Norman Borlaug to make this next revolution in agriculture work. From agriculture, contemporary heroes must emerge from at least four areas: plant breeding, irrigation technology, agronomy, value chain management. Nor should we ignore the role of sociologists, economists, and environmental biologists. Last, but not least, the role of consumers will be critical in keeping the process in line.

7.3.2.1   Plant Breeding

The goal for plant breeder heroes will be new crop varieties with the following characteristics:

– Nutrient uptake efficiency, for both macro- and micro-nutrients

– Pest resistance

– Improved photosynthetic assimilation rate

– Water uptake efficiency

– Resilience to stress

With the above traits, crops will need fewer inputs to be applied by the farmer, resulting in the potential for a more environmentally and economically sustainable system.

Fortunately for plant breeders, there have been major advances since the time of Norman Borlaug. One new technique is the evolving technology of genetic modification. In addition to the conventional methods available in the mid-twentieth century, new techniques of improving traits in varieties are part of the plant breeder toolbox. These include the evolving technology generally called genetic modification or engineering, which refers to the process of inserting an organism with a gene it does not naturally possess. In agricultural crops, the application of the methods for genetic modification has been able to improve crop protection, yield, quality, and nutritional composition. The technology presents other opportunities to use crop plants for producing high value-added molecules for pharmaceuticals, vitamins, or biopolymers for the industry.

Conventional plant breeding is a slow process, typically taking 12 years to breed a new crop variety. Biotechnology allows the breeding process to occur quickly, with methods that isolate specific traits in DNA and move them to another organism. In the past, this process might require multiple generations of crossbreeding to achieve the same result, if it were possible at all. GM biotechnology offers precise and specific genetic modifications in a short time period.

The most recent addition to the plant breeder toolbox is CRISPR technology. This acronym is short for "clusters of regularly interspaced short palindromic repeats" and it does not involve adding new genetic material to an organism. Rather, it is a new technology for editing existing genomes. The technique was developed by observations on the natural defense mechanisms of bacteria and single-celled microorganisms. It builds on the understanding of how those organisms respond to attacks by viruses, in which they actually destroy the DNA of the virus. From the practical observation of bacterial defense, the CRISPR technology has been developed to facilitate researchers changing DNA sequences and modifying gene function within an organism, without adding any new genetic material. Thus, technically the resultant changes are not GMO since no foreign genes have been added. This new technology is just now being adopted for fast and accurate crop breeding purposes. It stands to provide substantial opportunity for crop improvement, once fully developed.

There is considerable scientific and social discussion and dilemma around GMO. While necessary, these need to be resolved without undue delay. Scientific approaches and policy makers will need to regulate the product, rather than the process. Plant breeders will have to use the best combination of scientifically-based techniques to produce stable, cost-effective, and safe new varieties.

7.3.2.2 Irrigation

The so-called "elephant in the room" which will affect all initiatives to increase agricultural production is water. Globally, nearly 75% of the water consumed is used for agriculture. Irrigation is one of the key reasons we have been successful in feeding more than 7 billion souls. While only 20% of the world's farmland is irrigated, it produces 40% of our food supply. The highest yields obtained from irrigation are more than double the highest yields from rain-fed farming. Even low-input irrigation is more productive than high-input rain-fed farming. Add to this, the fact that humans today use 3X the water than we did 50 years ago. We need to continue to use irrigation to produce food but this overuse trend needs to be reversed. Engineers and irrigation technologists will need to develop techniques and adapt present technology in order to irrigate more land, not less, but do it smarter, with less water use. That water may be from nonconventional sources such as harvested rainwater, recycled wastewater, and desalinized waters. From the technology standpoint, improved on-farm water management means having accurate measures of soil water and plant stress monitoring. It will involve the precision application of water and will need to be combined with appropriate agronomic practices such as mulching and cover crops. Traditional husbandry methods have a lot to offer in terms of water conservation. This will take good communication between farmers and technologists, both in sharing information and in developing workable solutions.

7.3.2.3 Agronomy

The experience of traditional farmers combined with emerging knowledge from the new generation of organic farmers will help to develop agronomic strategies that are sustainable as part of overall production systems. Techniques such as composting, intercropping with legumes, and agroforestry are just a few approaches to be adopted. In a new sustainable green revolution, decreased reliance on synthetic chemicals will require intensive management, so that yields are maintained and increased. Efforts must be made to raise native fertility and to boost efficiency on existing cropland. We cannot continue to threaten vital rainforests and wetlands with agricultural exploitation. Crop management needs to be intensified, but sustainably.

7.3.2.4 Value Chain Management and Consumers

All those in the value chain and particularly we, as consumers, have a responsibility to at least halve food loss wastage. If we are to meet our goals of feeding our population in the future, it is essential to substantially decrease what is being lost. This will involve both technology and public policies in local, national, and international food value chains. In developed countries where much of the food waste is at the restaurant, food service, or home level, the onus is on the consumer to take the responsibility to decrease waste. Reducing the consumption of animal products can also improve the ability of the food system to supply the growing population.

Conclusion

The challenge for humanity is to plan and execute a new truly "green" Green Revolution in food production. The application of basic concepts of agroecology will help set the plan but the execution of the plan will rely on people throughout the value chain stepping forward to make it work.

Chapter 8
Wastes or Resources in Agroecosystems?

S. Wang and C. D. Caldwell

"If it can't be reduced, reused, repaired, rebuilt, refurbished, refinished, resold, recycled or composted, then it should be restricted, redesigned or removed from production".
— Pete Seeger

Abstract Resources and wastes continuously flow into and out of agroecosystems. Diversity is an important aspect of resources, especially when pertaining to genetic, soil and water resources. However, this chapter focuses mostly on the handling of wastes and furthermore how to turn wastes into resources. Using animal and plant manure as organic fertilizer adds fertility and structure to soil through the addition of nutrients, organic matter and bacteria. Biogas is a useful resource for both agricultural production and producing energy, acting as a low-cost fuel source. Finally, we discuss composting raw materials, outlining the process, factors affecting the process and techniques to improve aerobic composting. Composting offers huge benefits as a source of nutrients and waste control that ranges from home gardens to farmyards to large-scale industrial operations.

Learning Objectives
After studying this topic, students should be able to:

1. Differentiate between agroecosystem resources and wastes.
2. Identify the different composition and uses of animal and plant manures.
3. Discuss the uses of biogas.
4. Define the following terms:

   (a) Composting
   (b) Active composting
   (c) Compost
   (d) Feedstock
   (e) Thermophilic
   (f) Mesophilic
   (g) Curing (maturing)
   (h) Compost quality

(i) Vermicompost

5. Explain the process, factors affecting and techniques for managing compost.
6. Describe four agricultural uses of compost

8.1   What Are Resources and Wastes in Agroecosystems?

Agroecosystems are shaped by biotic and abiotic factors. These relatively open systems produce a flow of materials (outputs) for human use. In order to maintain ecological processes and an energy balance while simultaneously increasing the conversion efficiency of energy, an agroecosystem requires tremendous amounts of energy subsidies from society. From an energy flow analysis standpoint, wastes and by-products accompany every production process in agroecosystems by way of resource utilization (Fig. 8.1). Table 8.1 lists and categorizes the resources required for the running of an agroecosystem, including both biotic and abiotic factors and energy subsidies. These resources are categorized into physical, biological, socioeconomic and cultural inputs. The wastes and by-products of the production system are almost all the same (i.e. respiration gives off CO_2 and wastes, and CO_2 accompanies fossil fuel burning, Fig. 8.1).

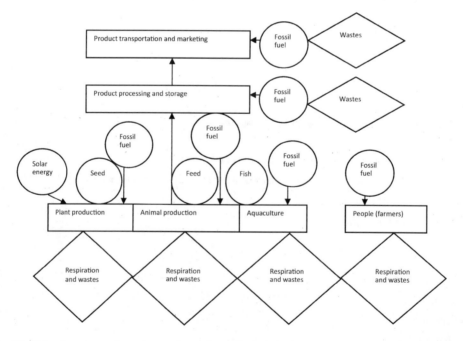

Fig. 8.1   Agroecosystem processes are dependent on various resources producing wastes

8.2   Resources in Agroecosystems

8.2.1   Genetic Resources

Genetic resources of crop plants and microorganisms play an integral role in sustainable agriculture. Development of improved agricultural crops and animal breeds through repeated natural and human selection over time results in a loss of genetic diversity. Unforeseen problems may include a reduced germplasm base and genetic fragility in crops. However, efforts to combat the loss of genetic resources can be seen all around the world. Plant breeders are accessing diverse genetic material from the centres of origin of our crop plants in order to stabilize the genetic base for our core crops.

Table 8.1   Resources in agroecosystems

Types of resources	Classification of resources
Physical	Radiation (solar energy)
	Temperature (heat)
	Rainfall, water supply (moisture)
	Soil condition
	Slope
	Land availability
Biological	Natural vegetation and their interactions
	Soil biota and their interactions
	Agrobiodiversity (crop, livestock and poultry and other domesticated microorganisms) and their interactions
Socioeconomic	Labour and draft forces
	Fertilizer and manure
	Pesticides, insecticide and herbicides
	Burning fossil energy
	Monetary capital
Cultural	Traditional knowledge
	Ideology in agroecosystem management
	Historical events

8.2.2　Soil Resources

Soil resources, another important aspect of sustainable agriculture, require good nutrient management. Improved soil and nutrient management techniques, attention to micronutrient levels and promoting processes such as production and use of compost help match the nutrient supply to the nutrient needs of the crop. The use of information technology and remote sensing in precision agriculture helps farmers make precise measurements of field conditions, respond to soil changes and provide inputs only where needed. Precision agriculture is useful both on a field scale and on a broad landscape scale. Therefore, it is applicable for large, capital intensive, individual farms in countries such as the USA but also on small farms around the world in response to environmental and crop needs.

8.2.3　Water Resources

Sustainable agriculture demands consistent, reliable availability of water resources and high efficiency in use of the water for production purposes. In many countries, water is a scarce resource, limiting the food supply available to those countries. Currently in high-income countries, little incentive exists for farmers to adopt watersaving strategies because the cost of water and the cost of wasting water is not born by the user. However, in poorer, more densely populated countries, irrigation is needed for agriculture and costs too much for poor farmers on small farms. Scientists and policymakers in many Asian countries realize that water availability will help alleviate poverty and malnutrition; therefore, policies to address water use and water availability are now becoming essential goals in national development. Government subsidies or reducing water fees for low-income users can give poor farmers an advantage to sustain themselves, which in the long term should cost less to the government and its people.

8.3　Waste in Agroecosystems

Ecologically, every production process produces wastes. In food production systems, these wastes are often organic based and vary from animal manures and postharvest wastes all the way to the spoiled food and unsold produce at supermarkets. However, recycling and reuse can change what was considered a waste into a resource and make the complete value chain more efficient and more economically and environmentally sustainable. Therefore, recycling and its environmental and economic impact is an essential component of managing agroecosystems. A good understanding of how wastes can be reduced and/or recycled into resources is key to sustainability.

Animal and Plant Manures

Manure is organic matter that is used to positively amend soil by improving soil structure, introducing nutrients (both macronutrients and micronutrients) and the stimulation of the living soil ecosystem through the addition of bacteria and fungi. There are three main classes of manures used in soil management: animal manure, plant manure and biogas.

1. Animal Manures (Brown Manures). Typically, animal manure refers to faeces used for organic fertilizer. Manure from different animals has different qualities, nutrient content and composition according to the type of animal, diet and maturity of the animal, and therefore different application rates are recommended. Farm slurry and farmyard manure refer to common forms of animal manure. Farm slurry, or liquid manure, occurs as a result of more intensive livestock rearing that uses added water as the medium for handling. Farmyard manure uses straw bedding to absorb faeces and urine and therefore will contain plant material and thus higher levels of organic matter. Many types of animal manures also contain other animal products such as wool, hair and feathers.

   Animal manure has been used for agricultural fertilizer for centuries. Animal dung adds organic matter which improves soil structure, increases the ability of the soil to hold nutrients and water, promotes soil microbial activity to improve plant nutrition and adds nutrients to assist in plant growth.

   Generally, farmers spread manures from chickens, pigs and cattle across a field with a manure spreader. However, manures from human sewage or intensive pig farming slurry may have particularly unpleasant odours. In these cases, where possible and the equipment is available, manure is knifed (injected) directly into the soil to reduce release of the odour. One other consideration for the farmer is to ensure that the manures are not in any way contaminated with medicines, chemicals or pathogens that may contaminate their soils or the nearby watercourses.

2. Plant Manures (Green Manures). Plant manure refers to crops grown for the sole purpose of increasing soil fertility through ploughing them into the soil, incorporating nutrients, stimulating microorganisms and increasing organic matter. Different plants incorporate different nutrients into the soil. Leguminous plants, such as clover, fix nitrogen using Rhizobia bacteria in specialized nodules in the root structure. The best green manures usually have extensive fibrous root systems that help in building up soil structure and soil health, as benefits beyond nutrient cycling. Deep roots of some green manures also work to extract nutrients up from lower levels of the soil back into the upper levels where they may be available for the next cropping season. They also provide large pore spaces for macroinvertebrates such as earthworms.

3. Biogas. The anaerobic (in the absence of oxygen) digestion of organic material can be used to

produce the biofuel known as biogas (also known as marsh gas). Biogas is principally composed of methane, a very combustible gas which can be used both for heating and cooking purposes on a small scale household level. On a larger scale, it may be used to run engines or to generate electrical power. The production of biogas can be a very useful by-product of waste plant biomass, animal wastes or any organic material such as food waste or even crops grown for energy. The production of biogas can be a key link for changing wastes into a resource.

Similar to natural gas, biogas can also be compressed and used to power vehicles; unlike natural gas, biogas is a renewable energy source and is a key resource in many parts of the world. For example, in India, an estimated two million households use appropriate level digestion facilities in the home to produce biogas. Usually these digesters are linked to animal production and the airtight digester pit digests manure from the adjacent animal unit. Because the technology can be applied simply with readily available, relatively inexpensive materials, the use of biogas can be a very environmentally sound approach. Beyond the production of the gas itself, the residual organic material left after the digestion is an excellent fertilizer.

8.4　Compost: Making It

8.4.1　What Is Compost?

Composting refers to the natural process of organic matter decomposing by micro-organisms under controlled conditions (FAO, 2004). The stabilized compost product serves as humus, a soil amendment or fertilizer (Fig. 8.2). In a farm setting, compost usually consists of a combination of food waste and animal waste undergoing aerobic decomposition.

Compost improves the physical, chemical and biological properties of soil, resulting in the following improvements: The soil ①becomes more resistant to stress (i.e. drought, disease and toxicity), ②improves the crop's ability to take up plant nutrients and ③has vigorous microbial activity, enhancing the active nutrient cycling capacity. These improvements ultimately help farmers reduce cropping risks, produce higher yields and spend less capital on inorganic fertilizers. Compost offers other on-farm benefits. Compost reduces waste mass, volume and odours, destroys pathogens, kills weed seeds, improves transportability and nutrient qualities, decreases pollutants, available for land application when convenient and acts as a saleable product. A few disadvantages of on-farm compost exist as well. Compost also increases the loss of ammonia, involves time and labour and requires some initial capital and operating costs for equipment and marketing for sale.

Fig. 8.2   A final compost product from a student group on the campus of Fujian Agriculture and Forestry University, 2007

8.4.2   The Composting Process

Composting is a controlled, managed biological oxidation process whereby microorganisms (bacteria, fungi) convert raw organic "wastes" into a stabilized organic residue called compost. Aerobic composting takes place in the presence of ample oxygen and occurs in two stages: an active stage and a curing stage. During the active stage, microorganisms feed on organic matter and consume oxygen while producing heat, carbon dioxide, ammonia and water vapour. A management plan for maintaining proper temperature, oxygen, moisture and other factors affecting the composting process is important in the active composting stage to achieve complete decomposition of all the biodegradable matter. During the curing stage, microbial activity slows down and nearly stops as the process reaches completion, resulting in a relatively stable, organic product (compost).

Composting processes range from extremely simple methods practised by individuals at home, semi-intensive methods practised by farmers on their land, to largescale operations practised industrially in cities. Composting systems vary from low technology to highly sophisticated, automated technology. Table 8.2 describes five basic composting methods: bins, passive windrows, active windrows, aerated static windrows and in-vessel channels. Composting has been considered primarily as a waste management strategy but, if done properly, it produces an excellent soil amendment. These two objectives complement each other.

Table 8.2　Basic composting methods (Manure Composting Manual, Government of Alberta, 2005)

	Bin	Passive windrow	Active windrow	Aerated static windrow	In-vessel channel
General	Low technology, medium quality	Low technology, quality problems	Active systems most common on farms	Effective for farm and municipal use	Large-scale systems for commercial applications
Labour	Medium labour required	Low labour required	Increases with aeration frequency and poor planning	System design and planning important. Monitoring needed	Requires consistent level of manage-ment/ product flow to be costefficient
Site	Limited land but requires a composting structure	Requires large land areas	Can require large land areas	Less land required given faster rates and effective pile volumes	Very limited land due to rapid rates and continuous operations
Bulking agent	Flexible	Less flexible, must be porous	Flexible	Less flexible, must be porous	Flexible
Active period	Range: 2–6 months	Range: 6–24 months	Range: 21–40 days	Range: 21–40 days	Range: 21–35 days
Curing	30+ days	Not applicable	30+ days	30+ days	30+ days
Size: Height	Dependent on bin design	1–4 metres	1–2.8 metres	3–4.5 metres	Dependent on bay design
Size: Width	Variable	3–7 metres	3–6 metres	Variable	Variable
Size: Length	Variable	Variable	Variable	Variable	Variable
Aeration system	Natural convection and mechanical turning	Natural convection only	Mechanical turning and natural convection	Forced positive/ negative airflow through pile	Extensive mechanical turning and aeration
Process control	Initial mix or layering and one turning	Initial mix only	Initial mix and turning	Initial mix, aeration, temperature and/or time control	Initial mix, aeration, temperature and/or time control and turning
Odour factors	Odour can occur, but generally during turning	Odour from the windrow will occur. The larger the windrow, the greater the odours	From surface area of wind-row. Turning can create odours during initial weeks	Odour can occur, but controls can be used, such as pile insulation and filters on air systems	Odour can occur. Often due to equipment failure or system design limitations

8.4.3　Factors Affecting Aerobic Composting

8.4.3.1　Aeration

By definition, aerobic composting requires the presence of oxygen, ideally at a concentration greater than 10%; lower oxygen concentrations or anaerobic conditions suppress aerobic microorganism growth and cause poor compost quality. Aeration throughout the raw material reduces odours, removes excess heat and encourages complete decomposition of carbon to release carbon dioxide rather than methane. Heat removal is particularly important in warm climates where the risk of overheating and fire exists.

8.4.3.2   Moisture

Moisture content supports the metabolic activity of microorganisms and indirectly supports the supply of oxygen. Composting materials should maintain moisture content between 45 and 65% by weight to ensure adequate moisture without limiting aeration. Bacterial activity slows down with moisture content below 40% and ceases entirely below 15%. Moisture content over 60% leads to leached nutrients, reduced porosity, increased odours and a slowed decomposition rate. In practice, desirable compost pile moisture content would start around 50%–60% and finish around 30%.

8.4.3.3   Nutrient Availability

The composting process requires adequate levels of carbon, nitrogen, phosphorus and potassium in order to feed the developing microorganisms. The microbes convert these nutrients from raw organic materials (e.g. manure, livestock mortalities) to stable forms with a lower likelihood of nutrient loss through volatilization and leaching. The C : N ratio is especially important among the required nutrients. An optimal C : N ratio of raw materials ranges from 25 : 1 to 30 : 1 with an acceptable range between 20 : 1 and 40 : 1 because microbes need at least 20–25 times more carbon than nitrogen to remain active. Microorganisms use carbon for energy and nitrogen for protein and reproduction. Ratios higher than 40 : 1 limit microorganism growth slowing the decomposition process. Ratios less than 20 : 1 underutilize nitrogen, potentially losing nitrogen to the atmosphere as nitrous oxide or ammonia. The stable end product compost should have a C : N ratio between 10 : 1 and 15 : 1.

8.4.3.4   Temperature

Temperature changes in composting piles indicate changes in microbial activity. To destroy pathogen viability, temperatures should remain above 55°C for a minimum of 2 weeks. To eliminate weed seeds, the temperature should remain above 62°C; however, microorganisms will begin to die around 70°C. During the active composting stage, higher temperatures result from vigorous microbial activity. During the early stages of the compost process, mesophilic (10–40℃) bacteria are active; as the pile becomes hotter, these bacteria die off and are followed by thermophilic bacteria (40–65℃). With each 10℃ increase in temperature, biochemical reaction rates nearly double. To regulate temperature and to maintain aeration, it is necessary to regularly turn compost piles.

8.4.3.5 pH Value

Adequate pH levels for microorganisms range between 6.5 and 7.5. Most animal manures pH ranges between 6.8 and 7.4 before major changes occur in the decomposition process. Lowered pH (higher acidity) may result from the temporary release of organic acids. Heightened pH (higher alkalinity) may result from the production of ammonia from nitrogenous compounds.

Of all the factors listed above, regulating temperature and pH are the most important (Fig. 8.3).

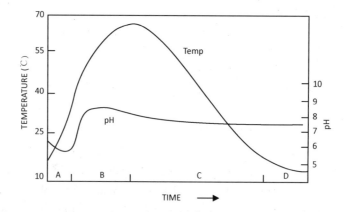

Fig. 8.3 Temperature and pH phases of composting

A. Mesophilic, 10–40℃; B. Thermophilic, 40–65℃; C. Cooling, 65℃ to ambient; D. Maturing (curing)

8.4.4 Techniques for Managing the Composting Process

Composting under different climatic and material conditions with varying biological, physical and chemical properties requires different technical applications. Every compost pile needs specific handling, whether the composition differs slightly or significantly. The following section describes techniques for effective aerobic composting; however, it is necessary to monitor compost piles to ensure that the base conditions for good composting are maintained.

8.4.4.1   Improved Aeration

As stated above, compost must receive enough oxygen to allow aerobic microorganisms to turn raw materials into a stable end product. Pile size, porosity of material, ventilation and turning practices all help improve compost aeration.

In high piles, anaerobic zones occur near the centre, slowing the overall composting process. In small piles, heat loss occurs quickly and sometimes fails to achieve adequate temperatures to kill pathogens and weed seeds. Porous materials should be kept in loose piles, limiting heavy weight on top in larger piles. Climate must also be taken into consideration; in cold climates, minimize heat loss by creating larger piles; in warm climates, reduce pile sizes to avoid overheating and potential fires.

Incorporating ventilation techniques complements efforts to optimize pile size. Ventilation holes supply more air to oxygen deficient areas of the pile, generally in the centre. The Chinese rural method of compost ventilation employs bamboo poles deep into the pile for 24 hours.

The technique of turning the compost pile results in a uniform distribution of air, prevention of overheating and killing of microbes. The Chinese rural composting pit method reduces a composting time of 6–8 months to 3 months by turning the pile three times throughout the maturing process. Worldwide, different methods include turning just once to turning once every day. However, too frequent turning may lower temperature and slow the composting process altogether. Fig. 8.4 demonstrates a small scale turning of the compost pile, while Fig. 8.5 shows the use of a commercial compost turner in a windrow system.

Fig. 8.4   Turning the compost pile at FAFU, 2007

Fig. 8.5　Turning a large-scale commercial compost windrow in Oregon, USA. This compost is used to fertilize a local orchard (photo credit: Washington State University extension service)

8.4.4.2　Inoculation

Some compost piles require inoculum organisms to enhance microbial activity, even after improving pile aeration. Inoculation with commercially available inoculants allows for rapid composting and composting of weeds.

8.4.4.3　Other Measures

Supplemental nutrition, shredding the pile and adding lime to compost piles are other methods that facilitate composting. Supplementing nitrogen, phosphorus, sugar or amino acids often complements the techniques mentioned above, boosting microorganisms' initial activities. Modifying a high C∶N ratio can significantly improve the composting process. Shredding the pile increases the surface area available for microbial action while providing better aeration. Adding lime weakens the lignin structure enhancing the microbial population but can also result in a higher pH value.

8.5   Vermicomposting

Another type of composting, called vermiculture, has the advantage that it can function at lower temperatures. As the term indicates, vermiculture is a type of compost that utilizes worms to facilitate the digestion of organic material into a usable, uniform organic product. Different types of worms may be utilized in this process, the most common of which is the red wigglers. The worms are introduced into the organic material and they ingest and digest the material, producing their natural excretion, which is essentially a very rich compost.

The rearing of worms in this way is a basic, easily mastered technology that can take various mixtures of organic material and produce excellent organic fertilizer. It can be done at the household level or may be expanded to larger operations to handle mixtures of manure and other farm wastes. This combines both appropriate technology and good environmental stewardship.

8.6   Agricultural Uses of Compost

The stabilized product from composting serves as a rich and important source of nutrients commonly used in modern agriculture. The organic material is used as a soil amendment, growing medium component, organic mulch, topsoil replacement or an amendment for soil blending.

8.6.1   Soil Amendment

Incorporating compost into the soil as an amendment may be used for topdressing crops such as turf. Principally, compost as a soil amendment is valued as a soil conditioner rather than a fertilizer. Most compost has a fertilizer ratio in the range of 1-1-1; that means it is 1% nitrogen, 1% phosphate and 1% potash—a very low analysis fertilizer. Typically, commercial fertilizers such as soluble chemical fertilizers have a much higher analysis, such as 10-10-10 or 10-15-30. Although compost is a very low analysis fertilizer, the value lies in its slow release of nutrients, its stimulation of microbial action and its organic matter composition. Nutrients in a slow release fertilizer are not very soluble and do not release rapidly; therefore, they do not leach easily. Soil may benefit from a slow release fertilizer by having the ability to retain more nutrients in the soil. A slow release fertilizer is very advantageous in certain situations, but it is unable to provide the same effects as quickly as most commercial fertilizers. However, compost has tremendous value as a soil conditioner; it improves the physical properties of the soil by improving the soil's water-holding capacity, tilth, water infiltration and nutrient holding capacity.

8.6.2　Growing Medium Component

Using compost as a growing medium, potting mix and potting soil is the second major use of compost in modern agriculture. This product may be bagged for commercial sale as a growing medium for plants. Compost provides a loose, friable (separate texture and structure), moisture retentive, high water-holding capacity, good growing medium for plants while suppressing or inhibiting plant diseases. Using compost to suppress diseases is particularly attractive to organic farmers, who (due to organic production regulations) cannot use inorganic fungicides, insecticides or herbicides, and has become a very active area of research. Compost, also used as organic mulch, can offer the benefits of suppressing weeds, conserving moisture and adding organic matter to the soil.

8.6.3　Topsoil Replacement/Amendment for Soil Blending

An increasingly popular use of compost employs compost as a component of topsoil replacement, a soil blending mix. Compost has a very high-volume market, as it can accept large amounts of compost around construction sites. Significant soil disruption occurs when building a house, restoring a dilapidated building or other industrial activities. Compost can help restore vegetation, a very difficult task, to those locations where soil has been stripped and removed.

Summary

Resources and wastes continuously flow into and out of agroecosystems. Diversity is an important aspect of resources, especially when pertaining to genetic, soil and water resources. However, this chapter focuses mostly on the handling of wastes and furthermore how to turn wastes into resources. Using animal and plant manure as organic fertilizer adds fertility and structure to soil through the addition of nutrients, organic matter and bacteria. Biogas is a useful resource for both agricultural production and producing energy, acting as a low-cost fuel source. Finally, we discuss composting raw materials, outlining the process, factors affecting the process and techniques to improve aerobic composting. Composting offers huge benefits as a source of nutrients and waste control that ranges from home gardens to farmyards to large-scale industrial operations.

Chapter 9
Global Climate Change and Agriculture

C. D. Caldwell and S. Smukler

"Reductions in suitable plant growing days will be most pronounced in tropical areas and in countries that are among the poorest and most highly dependent on plant-related goods and services."
—Camilo Mora

Abstract Agricultural practices and our food production and distribution system contribute significantly to greenhouse gas (GHG) emissions and therefore to the phenomenon of global climate change. The main GHG's from agriculture and food are carbon monoxide, methane, and nitrous oxide. This chapter addresses the principal concepts of the greenhouse effect and how agriculture can be a key component of both adaptation and mitigation of climate change.

Learning Objectives

After studying this topic, students should be able to:

1. Define the following terms:

   (a) Greenhouse effect
   (b) Carbon sequestration
   (c) Global climate change

2. Explain the relationship between atmospheric carbon dioxide concentration and global climate change.
3. Name two other important "greenhouse gases" and explain their significance.
4. Describe how agricultural practices may tend to increase the greenhouse effect.
5. Describe agricultural practices that may be used to increase stored (sequestered) carbon.
6. Explain how agriculture can mitigate greenhouse gases other than carbon dioxide.

9.1 What Is the Evidence for Climate Change?

9.1.1 Climate and Weather

Weather describes the short-term atmospheric properties such as wind, temperature, moisture, and atmospheric pressure at a time and place. Climate refers to long-term weather status. Climate changes result primarily from global air circulation patterns and solar energy capture. Air circulation is influenced by uneven heating of the earth's surface, tilt and rotation of the earth's axis. Climate change results from longterm variations in the amount of solar radiation reaching the earth, being captured and retained in the atmosphere, and the properties of the transfer of heat in air and water due to cyclical convection cells both in the atmosphere and in the oceans.

9.1.2　How Has the Global Temperature Changed?

Recent changes in temperature have varied with different parts of the earth. However, overall, the trend on a global basis has been for warming. Data from the 2019 NOAA global climate report indicate that when one combines the land and ocean temperature trends there has been a 0.07℃ increase every 10 years since 1880. The concerning situation is that there is an acceleration of this average warming rate of increase. Since 1981 the rate of increase has been 0.18℃.

9.1.3　What Impact Has Global Temperature Change Had on Climate?

Climate attribution is a scientific process for establishing the principal causes or physical explanation for observed climate conditions and phenomena (NOAA, 2018). It is very important for planning and strategy development, to maintain the scientific process for climate attribution and model prediction. Because climate change is so important for its potentially explosive impact on both economic and social fabric globally, before attributing man-made causes climate change, detailed statistical analysis is required. Over the last decade, hundreds of scientific papers have been published after rigorous peer review confirming the anthropogenic impacts on our climate. These papers and the science behind them are the basis for the predictive models that we now use to strategize for both adaptation and mitigation. Some of those documented impacts are as follows:

9.1.3.1　Arctic Ice

Changes in the nature of the sea ice in the Arctic support the conclusion regarding overall warming trends. If one compares the sea ice in the Arctic in 1985 with what we now see, it is apparent that there have been dramatic changes. The length of time that sea ice is present has decreased significantly over the past 35 years. Previously, approximately a third of the ice in the Arctic in March would be at least 4 years old; at present, just a little over 1% of the ice in March is of that age. In addition, 75% of the winter ice pack at present is only a few months old. This is a startling trend that shows the speed at which climate change is occurring (Fig. 9.1).

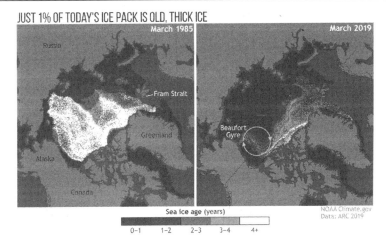

Fig. 9.1 **The age of sea ice in the Arctic ice pack during the third week of March 1985 (left) and March 2019 (right)** [Adapted from the 2019 Arctic Report Card (NOAA)]

Ice that is only one winter old is the darkest blue. Ice that has survived at least four full years is white

9.1.3.2 Wildfires

Two thousand and twenty was a year of extremes across the globe. The Southwest of Australia and parts of California have experienced ravaging wildfires, which caused extensive damage to the ecosystem and wildlife, and unfortunately, loss of human life. Wildfires are a natural part of the maintenance of many ecosystems; however human intervention, combined with climate change have resulted in extremes not known previously. Much of the world has experienced shifting of seasons, especially increased spring and summer temperatures, loss of snowpack, and moisture stress. This is resulting in much longer periods of sensitivity to severe wildfires.

9.1.3.3 Flash Floods and Droughts

The frequency and severity of floods are increasing across the world in response to climate change. These floods are occurring particularly more often in Asia and Africa but data from South America shows that in the years between 1995 and 2004, an average of 600,000 people was affected by floods; 10 years later that had increased to 2.2 million people and the numbers continue to rise every year due to flooding. There is the immediate disaster of the actual flood, but it is often followed by the effects flooding has on agriculture; i.e., damage to crops and animals meaning less food and increased malnutrition or starvation, especially in the poorest parts of the world. When too much rain falls in one part of the world, it is usually accompanied by too little rain in another part of the world. This has been the increasing trend over the last decades, in which we have seen droughts in every continent, often followed by wildfires.

9.1.3.4 Irregular Monsoons

Even where the rainfall has been within the annual range in amount, the disturbance of climate change has resulted in changed patterns in rainfall timing. In countries where farmers depend on the predictability of rains and the rainy season, this has become a major challenge resulting in low crop yields and socioeconomic impact.

9.2 What Is the Greenhouse Effect and How Is It Related to Climate Change?

9.2.1 The Greenhouse Effect Defined

Fig. 9.2 illustrates the general principle behind the natural phenomena that we call the "greenhouse effect". In this illustration, radiation from the sun, which is mostly short wavelength, can penetrate the glass of the greenhouse; some of this radiation is absorbed inside the greenhouse but much is reradiated as longwave length radiation. The glass in the greenhouse acts as a barrier to the escape of the long wavelength of reradiated energy and it is trapped. This causes heating within the greenhouse. Fig. 9.3 illustrates the natural greenhouse effect that has maintained a consistent temperature on the Earth over several thousands of years. It also illustrates the changes that have

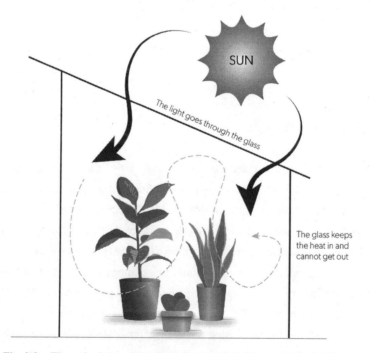

Fig. 9.2　The principles of the greenhouse effect (photo credit: S. Mantle)

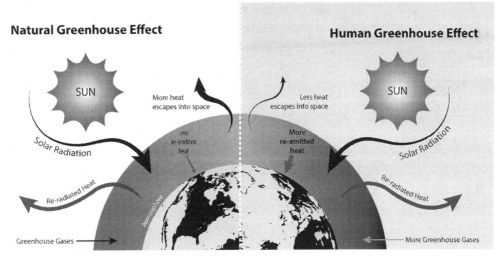

Fig. 9.3　Illustration of the human enhanced greenhouse effect (photo credit: S. Mantle)

occurred as we have increased the greenhouse gases in our atmosphere. These so-called "greenhouse gases" (GHG) act like the glass on the greenhouse, keeping much of the reradiated energy from leaving our atmosphere, causing the change in our global temperature. Climate scientists have demonstrated clearly that GHG concentrations have led to Global warming. Global warming refers to the earth's atmosphere warming due to increases in GHG concentrations, primarily from human activities. This overall global warming has resulted in significant variations and changes in climate across the globe. Because of this irregular effect of the overall warming, we now refer to this phenomenon as climate change brought about by human activities.

9.2.2　Relationship of Carbon Dioxide to Climate Change

Earth's average temperatures have naturally been changing throughout its history, and numerous research papers indicate a strong correlation between atmospheric CO_2 and earth's temperature. (Fig. 9.4) Research on atmospheric CO_2 measured from Antarctic ice cores has shown that the earth's climate had been cyclically stable over the past 400,000 years, but has experienced a rapid 30% increase in the past 250 years (1750–2000) from 280 ppmv CO_2 to 367 ppmv CO_2. In the last 60 years, the global CO_2 concentration has quickly climbed from 310 to 411 ppmv CO_2 (December 2019, NOAA) with an average global temperature elevated by 1.0℃ on the earth's surface. Carbon dioxide is one of the primary GHG's and has a major role in this change in global temperature.

9.2.3　Ocean Acidification

Oceans absorb about 25% of the CO_2 we release into the atmosphere every year, so as CO_2 levels increase, so do the levels in the ocean. This may appear to be a benefit as it originally was seen as the

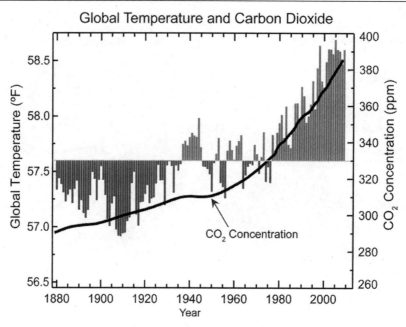

Fig. 9.4　Global surface temperature is rising

Red bars indicate temperatures above and blue bars indicate temperatures below the 1901–2000 average temperature. The black line

shows atmospheric carbon dioxide concentration in parts per million (NOAA)

ocean being a buffer against atmospheric increases. However, there is a downside—the CO_2 absorbed by the ocean is changing the chemistry of the seawater, a process called ocean acidification. Such acidification is threatening marine biodiversity especially as it is expressed in coral habitats.

9.2.4　Ocean Currents

Major ocean currents play a key role in distributing heat around the globe. There is a system of deep ocean circulation, driven by changes in the density of water affected by water temperature that is known as the global ocean conveyor belt. This system of ocean currents moves water around the globe, as it changes from cold dense water near the bottom and takes on heat to become warm less dense water at the surface. This worldwide water circulation process is a very important part of the global nutrient and carbon cycles. As the warm waters, which are naturally depleted of nutrients and carbon dioxide, travel through this mixing process, they are once again enriched. Nutrient-rich waters are the basis for the world's food chain which depends on the mixing of warmer waters with cooler waters, supporting the growth of algae and seaweed. The effects of changes in water temperature worldwide on this circulation process are being closely studied since any change, either in speed or pathway, would have a significant effect on the world's food chains and the local weather.

9.3　Other Greenhouse Gases

Since the 1750s, atmospheric GHG concentrations have sharply increased, mainly as a result of human activities. These activities include the widespread use of fossil fuels, deforestation, burning of grasslands, and the agricultural use of nitrogen fertilizers. Figs. 9.5, 9.6, and 9.7 show the recent increases in three of the key greenhouse gases affected particularly by agricultural practices: carbon dioxide, methane, and nitrous oxide.

Methane (CH_4) has 21 times the warming potential of carbon dioxide and an average atmospheric residence time of 7–10 years. The major worldwide sources of methane are from rice cultivation (due to anaerobic conditions in the paddy fields), livestock production (principally from ruminant animals and manure), decay from landfills, and mining operations. Agriculture contributes globally 50% of the methane released to the atmosphere.

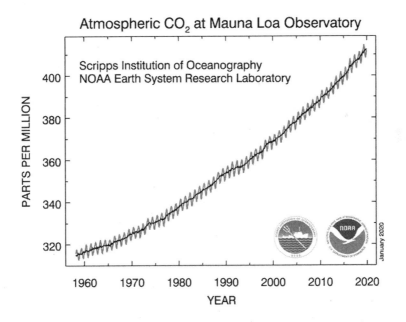

Fig. 9.5　Increase in atmospheric CO_2 from 1960–2020 (NOAA)

Nitrous oxide (N_2O) has 310 times the warming potential of carbon dioxide and an average atmospheric residence time of 140–190 years. The principal sources are from industry and agriculture, especially the production and use of inorganic nitrogen-based fertilizers. Agriculture contributes globally 70%–80% of the nitrous oxide released to the atmosphere.

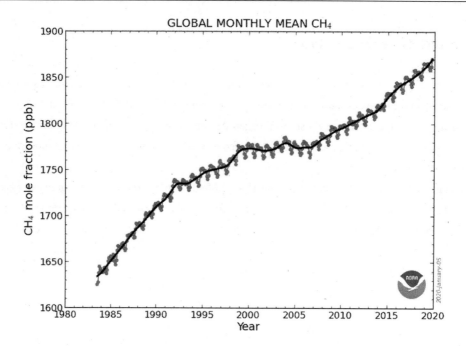

Fig. 9.6　Increase in atmospheric methane from 1960–2020 (NOAA)

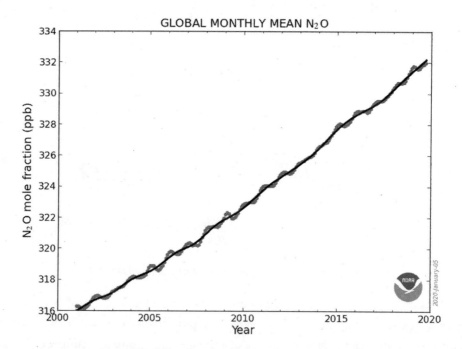

Fig. 9.7　Increase in atmospheric nitrous oxide from 1960–2020 (NOAA)

9.4   Agriculture's Contribution to the Greenhouse Gases and Global Climate Change

Agriculture and climate change are interconnected; climate change will influence agricultural crop production while agriculture continues to contribute to environmental change through emissions of greenhouse gases and altered land use. According to the Intergovernmental Panel on Climate Change (IPCC), the three main causes of increased greenhouse gases over the past 250 years were fossil fuel use, changes in land use, and agricultural production practices.

1. Land use: three types of agricultural land use contribute to increased GHG's:

   (a) Net deforestation releasing carbon dioxide

   (b) Rice cultivation releasing methane

   (c) Fertilizer application releasing nitrous oxide

2. Livestock: livestock production covers 70% of all land used for agriculture and contributes considerably to land use effects, since crops such as corn and alfalfa are grown as livestock feed. 18% of human-made GHG emissions come from livestock and livestock-related activities.

3. Cultivation leading to soil carbon losses.

Soils are inextricably connected to climate change as they currently account for most of the terrestrial organic carbon. In the top 2 m of soil, it is estimated that there is 3012 Pg Carbon (Sanderman et al., 2017), more than three times that found in the atmosphere (~830 Pg C). Thus any change in the amount of carbon in the soil can have an important impact on the climate (Ciais et al., 2013; Paustian et al., 2016). Carbon in soils has been incorporated through biological processes over millennia and is now playing an important role in how we deal with a changing climate. As plants grow, they utilize CO_2 from the atmosphere to build carbohydrates through photosynthesis. These carbohydrates are then deposited on the soil as plant matter (e.g., leaf litter) or into the soil as root exudates which can be used as an energy source by soil microorganisms. Once carbon from plants is broken down and incorporated into the soil it can be stored as soil organic carbon (SOC). SOC is generally classified into particulate organic matter (POM) which can be stored in the soil for <10 years to decades or, as more recalcitrant mineral-associated organic matter (MAOM), which can be stored for decades to centuries (Lavallee et al., 2020). While natural ecosystems maintain a relatively stable amount of carbon in the soil, this can change rapidly when managed for human use.

Since humans began cultivating the soil for agricultural production some 8000–10,000 years ago, it has been estimated that ~133 Pg C has been lost from the top 2 m of soil to the atmosphere (Sanderman et al., 2017). When soils are disturbed, for example by tilling to prepare for crop production, soil microorganisms have increased access to oxygen, which enables them to utilize the carbon stored in the soil as an energy source. As the microorganisms use carbon and oxygen as energy, they respire CO_2 into the atmosphere. Thus, when forests, grasslands, or other natural ecosystems are converted to agricultural production, large quantities of carbon can be lost to the atmosphere rapidly. Furthermore, when croplands, grasslands, and forests are managed for productivity by adding fertilizers that include nitrogen, microbial decomposition has been shown in some cases to be increased (Averill and Waring, 2018). Given the importance of carbon to the maintenance of soil quality and health (see Chap. 4), continuous management that leads to decomposition without adequately replacing the carbon lost has resulted in the degradation of soil and loss of functions around the world (FAO and ITPS, 2015).

9.5　Impacts of Global Climate Change on Agriculture

9.5.1　Global Winners and Losers

Mora et al. (2015) examined several projections of climate impacts using various mitigation scenarios in an attempt to define how the changes in the environment would limit crop growth and therefore have an influence on agroecosystems and the social systems that depend on them. As one would expect, all the models show increases in overall global temperature and growing days above freezing. However, surprisingly, in most cases around the world, suitable days for growing crops decrease by as much as 11% when other factors such as maximum temperatures, water availability, and solar radiation are considered. There are areas in the Northern Hemisphere, including much of Russia, Northern China, and Canada that are projected to increase their number of appropriate growing days; however, the models predict that much of the rest of the world will see decreases in suitable days for crop growth. This is particularly notable in tropical areas. These models are consistent with others (e.g., Ricke et al., 2018) in their conclusion that the effects of climate change will be uneven across the globe, both on a climate basis and a socialeconomic basis. Models predict that the poorest and most vulnerable people, which make up about 30% of the world's population, live in areas that will be affected the most by climate change. Impacts on suitable conditions for agriculture are projected to be less severe if we are successful in our more substantial mitigation strategies. It is obvious both for humanitarian and world security basis, that there is a need for immediate efforts at mitigation and adaptation.

9.5.2　Mitigation and Adaptation

Mitigation refers to those response strategies that reduce the sources of GHG's and adaptation involves the actions of adjusting practices to the threat of climate change. At this point, it is obvious that we need to do both in agriculture.

9.5.2.1   Carbon Dioxide

Agricultural managers are working towards a low carbon type of agricultural production system. This involves reducing fossil fuel use and substituting fossil fuel with biofuel in some situations. Increasing production efficiency must be linked with increasing overall value chain efficiency, lowering food losses, and decreasing food transportation and trade distances through appropriate localized food production systems and strategies.

9.5.2.2   Methane

Virtually all the CH_4 emission on temperate climate farms is from livestock. Ruminant animals release methane as part of the digestive process in the rumen. The waste products from animals (manure) also is a significant source. Emissions in tropical and sub-tropical countries are from a combination of the anaerobic soil conditions in rice paddies and livestock emissions. Fig. 9.8 compares the relative amounts of methane production from different livestock types. It is obvious from this figure, that one way to reduce the impact of agricultural systems on greenhouse gas production, is to decrease the numbers of ruminant animals and our reliance on beef, dairy, and dairy products. Maintaining some lower level of ruminant animals will require a change in rations to reduce the digestion time of ruminant animals, thus reducing methane production. Using solid manure systems as opposed to liquid manure systems also decreases the release of methane to the atmosphere. Manure should also be applied to land as quickly as possible after excretion and incorporated so that the methane is captured in the soil. Long-term storage of manure should be avoided and the amount of bedding in manure should be minimal. If it is necessary to store manure, there are methods to capture the methane for use as an energy source before it can be released into the atmosphere. Manure that is to be composted, should be aerated frequently to prevent anaerobic conditions.

9.5.2.3   Nitrous Oxide

Fig. 9.9 demonstrates the release of nitrous oxide as part of the nitrogen cycle in an agricultural production system. Nitrous oxide occurs at low concentrations (0.3 ppm) but has recently been increasing at 0.3% per year. Most N_2O from agriculture is produced in the soil and originates from two places in the nitrogen cycle, nitrification (converting NH_4^+ to NO_3^-) and denitrification (converting NO_3^- to gaseous N_2). Both processes are carried out by soil bacteria. Emissions from agricultural soils result from excessive nitrogen inputs and a disruption of the nitrogen cycle. Agricultural soils often have higher rates of N_2O than the natural vegetation; e.g., a fertilized rice field may release 5 kg N per ha per year. while a forest soil would emit 0.04 kg N per ha per year.

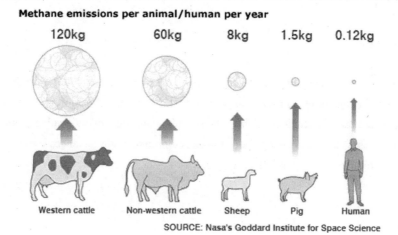

Fig. 9.8　A comparison of the methane emissions from different animal sources

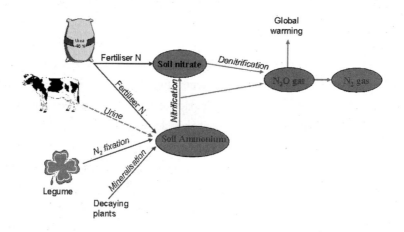

Fig. 9.9　Nitrous oxide release in the nitrogen cycle in agricultural systems

9.5.2.4　Soil Mitigation

Mitigation techniques for reducing nitrous oxide production in agricultural systems need to involve matching fertilizer applications more closely with crop needs so that there is not excess available for breakdown by soil bacteria. This involves farmers having well-developed nutrient management plans which include such management as avoiding excessive manure applications, optimizing the timing of any nitrogen application, and using effective, precise placement of fertilizers. The use of cover crops helps ensure that any excess nitrogen is tied up in organic matter and not released to the atmosphere.

Soil organic carbon losses are concerning as SOC may play an increasingly important role in helping land managers adapt to the expected impacts of climate change. Shifting precipitation patterns can result in waterlogging and have negative impacts on plant productivity or a farmer's ability to prepare their crops without damaging their soil. SOC, as a primary component of soil organic matter (SOM), can improve soil aggregation, help infiltrate and drain water and reduce its susceptibility to compaction. Alternatively, as the duration and severity of drought are expected to increase in some parts of the world, increasing SOM could improve soil water holding capacity and provide plants with access to water during these stressful times. Carbon stocks can be increased in soils by either reducing the rate of decomposition or increasing the amount of organic inputs (Paustian et al., 2016).

Increasing soil carbon sequestration can be achieved by several land management options. In situations where soils are degraded, left bare or fallow, returning them to perennial cover, using cover crops, adding nutrients by planting nitrogen-fixing species, or applying fertilizer and amendment can increase plant productivity and carbon inputs into the soil and reduce losses. Losses can be further reduced by limiting disturbance to the soil by minimizing tillage and retaining crop residues to cover the soil. Another way to increase carbon inputs would be to add compost or biochar frequently to the soil. If wetlands, or organic soils have been drained, returning them to their prior water regime reduces oxygen availability that would contribute to further oxidation and enables the restoration of plants that are adapted to saturated conditions. In situations where excess nitrogen is being lost to the environment and resulting in GHG emissions, application rates can be adjusted to better match plant needs in terms of the overall amount, timing, and location. Finally, agroforestry, planting trees in and around crop or pasture fields, also shows great potential for increasing SOC (Paustian et al., 2016).

Increasing SOC can not only help with climate adaptation but also mitigate the problem. It has been estimated that the potential of soils around the world to sequester atmospheric carbon ranges from 2 to 3 Gt C per year (Minasny et al., 2017). While soils can only sequester CO_2 to a certain point after which they reach an equilibrium, if practices were adopted to maximize sequestration, 20%–35% of global anthropogenic GHG emissions could be offset for the next 20–30 years (Minasny et al., 2017).

Although there is a well-established relationship between temperature and decomposition rates, the science is unclear as to how climate change will impact carbon stocks in the future. There is evidence that increased temperatures will result in a negative feedback loop, where temperature increases contribute to decomposition rates, resulting in increased losses of carbon from soil to the atmosphere. Global-scale analyses have suggested that, given no change in current trajectories of emissions, the expected increase of 2℃ over the next 35 years could result in carbon losses from the upper layer of soil of approximately 55 ± 50 Pg C (Crowther et al., 2016). This amount of SOC loss would increase atmospheric CO_2 concentrations by 25 parts per million (Crowther et al., 2016). Alternatively, modeling soil carbon stocks that account for expected interactions of temperature and moisture on plant productivity have shown the potential to gain as much as 253 Pg C by the end of the twenty-first century as a result of increased primary productivity (Todd-Brown et al., 2014). It however, remains unclear how the interaction of temperature, precipitation, and elevated CO_2 concentrations will impact soil carbon inputs from primary productivity and thus relative losses and this potential feedback loop.

Chapter 10
Agrobiodiversity and Agroecosystem Stability

S. Wang

"The best and most efficient pharmacy is within your own system."
—*Robert C. Peale*

Abstract Our living planet is uniquely full of living organisms, jointly forming the biodiversity that is the fabric of life. Biodiversity refers to the assemblage of all the species of plants, animals and microorganisms living and interacting within species, between species and in ecosystems. Biodiversity loss joins a litany of anthropogenic interdependent ecological 'crises', such as global warming; ozone layer depletion; acid rain; disappearance of tropical rainforests; dwindling varieties of wildlife; air, water and soil pollution and desertification. Pollution, rapidly growing in developing countries, promises to act as a catalyst for biodiversity loss on a catastrophic level. Agroecosystems depend on the presence of biodiversity for internal functions. However, many human activities such as the indiscriminate killing of wildlife and destruction of habitat have resulted in the loss of precious gene pools and a reduction in ecosystem services such as food production, nutrient cycling, microclimate and hydrological regulation, suppression of diseases and detoxifying of chemicals. Threats to agroecosystem biodiversity resulting from our present agriculture production systems require increased attention and actions to avert and mitigate the collapse of biological complexity. There are several ways to rebuild agrobiodiversity in agroecosystems by means of managing field, soil and crops.

Learning Objectives

After studying this topic, students should be able to:

1. List and describe the three main classifications of biodiversity.
2. Compare biodiversity between natural ecosystems and agroecosystems.
3. List the problems and causes of agrobiodiversity loss.
4. Explain the key principles of managing agroecosystems for maximum biodiversity stability.
5. Describe ways to enhance agroecosystem stability and increase biodiversity.

10.1 Biodiversity and Its Associated Ecosystem Services

Biodiversity refers to the assemblage of all species of plants, animals and microorganisms living and interacting within ecosystems. In 1992, the Earth submit held in Rio de Janeiro formally defined biodiversity (Alkorta et al., 2004):

> The variability among living organisms from all sources including internal, terrestrial, marine and other aquatic ecosystems and the ecological complexes of which they are a part; this includes diversity within species, between species and within ecosystems.

Scientists estimate that between 30 and 50 million different species inhabit the Earth (Paoletti et al. 1992), with nearly 14 million species scientifically classified and described (Wilson, 1988). While five major mass extinctions of species have occurred in Earth's history, recent extinctions are inferred to be human induced. Hence, it is urgent to take actions to conserve biodiversity under the framework of the Convention on Biological Diversity (CBD) by concentrating on "hotspots", areas with exceptional concentrations of endemic species and high rates of habitat loss since 1992 (Myers et al., 2000). To do this, a combination of the concept of ecosystem services and economy or market-based instrument has been proposed to better conserve biodiversity (Rodríguez-Labajos and Martínez-Alier, 2013).

10.1.1　Classifications of Biodiversity

At the "Convention on Biological Diversity" (CBD), issued at the Earth Summit in 1992 (Alkorta et al., 2004), biodiversity was classified officially as diversity at three levels: genetic, species and ecosystem (habitat). In practice, biodiversity conservation measures generally focus on an individual species in a natural or semi-natural setting.

10.1.1.1　Genetic Biodiversity

Genes determine which traits an organism exhibits; genetic biodiversity refers to all the genetic information contained in every plant, animal and microorganism on Earth. Genetic biodiversity helps ensure a species can evolve or respond to stresses such as diseases, predators, parasites, pollution and climate change (Kearns, 2010). For instance, many agricultural crops grow in monocultures, genetically homogenous species. While this practice usually consists of high-yield varieties that can be economically valuable, a single pest, disease or environmental disruption can ravage the entire crop.

10.1.1.2　Species Biodiversity

Species biodiversity refers to all the variety of plants, animals and microorganism species on the

Earth, including the tiniest single-celled microbes to the largest organism on the Earth, the blue whale. The greatest species diversity can be found among prokaryotic organisms (*Eubacteria* and *Archaea*). Prokaryotic organisms provide our environment with a multitude of functions, deriving energy from both organic and inorganic chemicals; recent discoveries show that many can survive in extreme conditions. However, even with an estimated 15,000 new species (big and small) described each year, scientists believe we are losing organisms at a faster rate than we can identify them.

10.1.1.3   Ecosystem Biodiversity

Ecosystem biodiversity relates to the variety of habitats, biotic communities and ecological processes that influence morphology, behaviour and interactions among species in an ecosystem (MEA, 2005). Pristine ecosystems preserve and support systems and biological processes that maintain life; these processes include nutrient and water cycling, photosynthesis, energy flow through the food web and patterns of plant succession (Kearns, 2010).

10.1.2   *Biodiversity Linking to Ecosystem Services*

Although ecosystem services (EEs) have been linked to human well-being since EEs were proposed and assessed (Ehrlich and Mooney, 1983; Costanza et al., 1997; MEA, 2005), biodiversity actually locates at the bridge point between these two concepts, making such nexus visible (Ring et al., 2010), because it is the biodiversity in an ecosystem which provides ecosystem services such as food, medicine, goods or environmental protection and other regulating and cultural services. Biodiversity conservation measures usually aim to increase ecological resilience, biological control and cultural or traditional values based on intrinsic or aesthetic values. While intrinsic values for biodiversity exist, most of the reasons we value biodiversity are homocentric. Biodiversity provides us with (MEA, 2005):

1. food, fuel, fibre, shelter and building materials.
2. purification of air and water.
3. detoxification and removal of wastes.
4. stabilization and moderation of climate, floods, droughts, temperature extremes and wind forces.
5. generation and recycling of soil fertility and nutrient cycling.
6. pollination of plants and crops.
7. control over pests and diseases.
8. maintenance of genetic resources and the ability to adapt to change.
9. cultural, aesthetic and spiritual values.

Furthermore, future gene adaptations may provide high value crops; biota may contain cures for emerging diseases and ecosystem services can be accessed for changing human needs (MEA, 2005). The current loss of ecosystems, species and gene pools can lead to the loss of critical ecosystem services.

Valuing biodiversity by linking it to EES is the first step to use economy or market-based instruments to help biodiversity conservation (Jones-Walters and Mulder, 2009). According to Zedan (2005), biodiversity directly accounts for 40% of the world economy estimated at US $18–61 trillion per year (Table 10.1). The ecological value of biodiversity far exceeds its economic market value. Accounting for just 17 ecosystem services for 16 biomes of the Earth's ecosystem, the minimum estimated value equalled approximately $33 trillion per year (Costanza et al., 1997). The annual market for products derived from genetic resources, including pharmaceuticals, botanical medicines, agricultural produce, ornamental horticultural products, crop protection products, biotechnologies and cosmetic products is between US $500–800 billion (European Commission, 2006). Likewise, the total estimated value of ecological services from biodiversity in China is more than 37.1×10^{12} Yuan RMB, far more than its traditional economic value of 1.72×10^{12} Yuan RMB in 1998 (Table 10.2; China's Biodiversity Status Research Group, 1998).

Table 10.1　Putting a price on biodiversity goods and services (Zedan, 2005)

Ecological functions	Economic values (USD)
Ecosystem services worldwide	$18–61 trillion per year
Soil microbial services	$33 billon per year
Global benefits from coral reefs	$30 billion per year
Insect pollination of over 40 commercial crops in the USA	$30 billion per year
Sales of prescription drugs containing ingredients from wild plants (in the USA)	$15 billion in 1990
Genetic traits from wild crop varieties introduced into domestic agricultural crops (in the USA)	$8 billion per year
Total seed-sector activities worldwide	$45 billion per year
Global market for herbal drugs	$47 billion in 2000

Table 10.2　Values of biodiversity in China (China's Biodiversity Status Research Group, 1998)

Value classified	Value types	Economic values ($\times 10^{12}$ Yuan RMB)
Direct use value	Marketing products	1.02
	Other direct used	0.78
Sum of subclass		1.80
Indirect use value	Organic matter enhancement	23.3
	CO_2 sequestration	3.27
	O_2 release	3.11
	Nutrient cycle and deposit	0.32
	Soil conservation	6.64

(continued)

Value classified	Value types	Economic values ($\times 10^{12}$ Yuan RMB)
Indirect use value	Water preservation	0.27
	Pollutant absorption	0.40
Sum of subclass		37.31
Potential used value	Selecting used value	0.09
	Preserving value	0.13
Sum of subclass		0.22
Sum total		39.33

10.1.3   Threats to Biodiversity

While global biodiversity includes millions of species, they are unevenly distributed in terms of location and threats.

Conservative estimates suggest we lose 5000 species a year, while less conservative estimates raise that number to 150,000 species extinctions per year (Goodland, 1992). Tropical rainforests and coral reefs are the pinnacles of biodiversity, 90% of described terrestrial species live in rainforests and 34%–53% of total described species live in coral reefs, while they only cover 5% of the area of global rainforests (Francini-Filho et al., 2018). Human activities have accelerated species extinction rate 1000 fold by eroding environmental services, responding only to crisis at the local and national level and spending 90% of conservation funding in economically rich countries that are biodiversity minimal (Brooks, 2012).

Recognition of decreasing biodiversity has increased in recent years; however, the rate of biodiversity loss has not declined. For example, even with the widespread public awareness over global deforestation, the world lost nearly 200 million hectares of forested land between 1980 and 1995. Biodiversity loss goes far beyond deforestation. Biodiversity is currently depreciating worldwide from the degradation of tropical rainforests, coral reefs, wetlands, grasslands and soil. Closely related problems include habitat destruction and invasive species outcompeting native species. Atmospheric pollution will only accelerate these global extinction rates.

Biodiversity loss combined with the loss of traditional knowledge will limit options for human adaptation to a changing environment. Decreased biodiversity will limit nutrition, food production and job opportunities.

Conservation Measures Partnership of The International Union for Conservation of Nature (IUCN-CMP) classified direct threats to biodiversity into 11 categories, beginning with the highest level of threat (Salafsky et al., 2008; Wong, 2012):

1. residential and commercial development.

2. agriculture and aquaculture.

3. energy production and mining.

4. transportation and service corridors.

5. biological resource use.

6. human intrusions and disturbance.

7. natural system modifications.

8. invasive and other problematic species and genes.

9. pollution.

10. geological events.

11. climate change and severe weather.

10.2　Biodiversity in Agroecosystems

10.2.1　Agrobiodiversity Loss

Agriculture has been a major contributor to biodiversity loss through monocropping, species domestication, selective breeding and hybridization. Replacing natural biodiversity with a limited number of cultivated plant varieties and domesticated animals in modern agriculture has simplified the structure of the environment. On the 1440 million hectares of cultivated land around the world, around 70 crop species are planted (approximately 12 grain species, 23 vegetable species, 35 fruit and nut species), whereas just 1 ha of tropical rainforest typically contains 100 species of trees alone (Altieri, 1999). In particular, agriculture has become increasingly dependent on a few bean, maize, wheat, corn, rice and cotton varieties. Tables 10.3 and 10.4 show the extent of genetic loss in crops and the reduction in diversity of fruits and vegetables. Scientific research has expressed concern time and time again over the great danger associated with genetic uniformity. While industrial agriculture and the Green Revolution have significantly increased global food production, with it came significant biophysical and socio-economic costs and disadvantages to many parts of the world.

Table 10.3　Extent of genetic loss in selected crops (Thrupp, 2000)

Crop	Country	Number of varieties loss
Rice	Sri Lanka	From 2000 in 1959 to fewer than 100 today, 75% descend from a common stock
Rice	Bangladesh	62% descend from a common stock
Rice	Indonesia	74% descend from a common stock
Wheat	USA	50% of crop in nine varieties
Potatoes	USA	75% of crop in four varieties
Soybean	USA	50% of crop in six varieties

Table 10.4   Reduction of diversity in fruit and vegetables 1903–1983 (Thrupp, 2000)

Crop	1903 (#)	1983 (#)	Loss (%)
Asparagus	46	1	97.8
Bean	578	32	94.5
Beet	288	17	94.1
Carrot	287	21	92.7
Leet	39	5	87.2
Lettuce	497	36	92.8
Onion	357	21	94.1
Parsnip	75	5	93.3
Pea	408	25	93.9
Radish	463	27	94.2
Spinach	109	7	93.6
Squash	341	40	88.3
Turnip	237	24	89.9

10.2.2   Problems Increasing Agrobiodiversity Loss

The threat of erosion of agrobiodiversity has been manifested in many ways within farming systems, off farms, in natural habitats and throughout communities worldwide. These threats derive from conflicting policies and inappropriate production practices. For example, livestock suffers from genetic erosion; as farmers concentrate on new breeds of chickens, sheep, pigs and cattle, traditional strains disappear. It is estimated that globally every week at least one traditional breed of livestock disappears, 16% of breeds of cattle, water buffalo, goats, pigs, sheep, horse and moneys have become extinct and another 15% are rare (FAO, 2019). These losses weaken breeding programs and risk the ability of livestock to adapt in the future. As consumers of animal products, we also lose nutrition, knowledge and cultural diversity.

10.2.3   Underlying Causes of Agrobiodiversity Loss

The proximate causes to the erosion of agrobiodiversity are tied to unsustainable technologies, land-degrading practices and overuse of chemicals in agriculture. Yet, these proximate causes have developed in close association to underlying ideologies, policies and education, business, demographic and socio-economic pressures (Thrupp, 2000). The relationship among the problems, proximate causes and underlying causes to agrobiodiversity loss can be seen in Table 10.5.

Agricultural systems can be classified by four parameters: biological diversity, intensity of human management, net energy balance and management responsibility. If we think of ecosystem management as a continuum, wilderness is on one end of the spectrum and intensively managed ecosystems on the other (Tivy, 1990, 2014). Agroecosystems range within this continuum, but try to closely adapt to the wild, ecological conditions with high biodiversity and lower management inputs. Tivy (1990) stated that in the intensively managed systems, "man's technical expertise is such that the physical environment is no longer a significant variable in determining or influencing the type of agroecosystem". The typical consequence of agrobiodiversity loss is exemplified by "The Great Irish Potato Famine" in the following knowledge box (Box 10.1).

Table 10.5 Addressing problems and causes of agrobiodiversity loss linked to agriculture (Thrupp, 2000)

Problems	Proximate causes	Underlying causes of all problems
Erosion of genetic resources (livestock and crops/plants) – threatens food security – increases risk –prevents future discoveries	Dominance of uniform high-yield varieties (HYVs) and monoculture, biases in breeding methods, weak conservation efforts	Industrial/Green Revolution paradigm that stresses uniform monoculture
		Inequitable distribution of land and resources
Erosion of insect diversity – increases susceptibility – ruins pollination and biocontrol	Heavy use of pesticides, use of monoculture/uniform species, degrading habitats, harbouring insects	Policies that support uniform HYVs and chemicals
		(e.g. subsidies, credit policies and market standards)
Erosion of soil diversity – leads to fertility loss – reduces productivity	Heavy use of agrochemicals, degrading tillage practices, use of monoculture	Pressures and influence of seed/ agrochemical companies and extension systems
Loss of habitat diversity including wild crop relatives	Intensification in marginal lands, drift/contamination from chemicals	Trade liberalization and market expansion policies that neglect social and ecological factors
Loss of indigenous methods and knowledge of biodiversity	Spread of uniform "modern" varieties and technologies	Lack of awareness of agroecology in R&D and in education institutions
		Disrespect for local knowledge
		Demographic pressures

Box 10.1   The Great Irish Potato Famine: A CASE Study in Biodiversity

The Great Famine in Ireland was a natural catastrophe caused by a potato disease between 1845 and 1852. Over that time period, the population of Ireland reduced by 20%–25%. Approximately 1 million people died of starvation and epidemic disease, and another 2 million people emigrated.

The potato was initially introduced to the gentry of Ireland as a garden crop, became a widespread supplementary food by the late seventeenth century and a staple of the lower societal classes by the early eighteenth century. In fact, a third of the population depended on potatoes for their entire sustenance.

As potato acreage expanded, the variety of potato species did not increase and diversify. Potato blight, caused by the fungus *Phytophthora infestans*, attacked in successive blasts ravishing most of the potato population for four to five consecutive years. Although the exact origin of blight is unknown, it was probably introduced through cargo ships, coming from Peru,

Baltimore, Philadelphia or New York. Between 1728 and 1851, the Census of Ireland Commissioners recorded 24 widespread crop potato failures. Although actions and inactionsof the Whig government, specifically concerning food security and housing regulations, intensified the magnitude of the situation, agricultural practices that include greater biodiversity may have ameliorated the situation or prevented the extensive ravishing of Irish farmland. The effects of the Great Famine have permanently shaped the demographic, political and cultural landscape in Ireland.

10.3　Regenerating Biodiversity in Agroecosystems

Agroecosystem biodiversity can be categorized into three intermingled and strongly interacting sub-systems: managed fields (productive sub-system), semi-natural habitats surrounding those fields and the human sub-systems consisting of settlements and infrastructures (Moonen and Barberi, 2008). Conservation typically focuses on the semi-natural sub-system to increase regional species pools and genome diversity, habitat diversity at all levels and to serve as a buffer against large-scale pest invasion and increase multi-functionality of direct economic returns. Recently, projects have considered the possibility of using agroecosystems to remediate environmental pollution from industrial activities to serve the purposes stated above.

10.3.1　Enhancing Agroecosystem Stability

Enhancing biodiversity in agricultural systems will play a key role in creating sustainable agroe-cosystems. Perhaps we should examine farming practices in third world countries, where agricultural land is marginal and therefore, practices are riskier. Traditional multiple cropping systems, traditional agroforestry systems, intercropping, shifting cultivation and other traditional farming methods generally mimic natural ecological processes, promoting species richness, production stability, efficient labour use and reduced pest and disease incidences while offering a diversified diet and maximum economic returns with little technological input. Traditional, multiple cropping systems alone can provide up to 15%–20% of the world's food supply (Altieri, 1999). Fig. 10.1 describes the integration of resources, components and functions for multiple-use farm systems which all help regenerate biodiversity in agroecosystems.

Agroecosystems benefit from resources in their natural habitats. According to Thrupp (2000), agrobiodiversity encompasses the following biological resources:

- genetic resources
- edible plants and crops
- livestock and freshwater fish
- soil organisms

- insects, bacteria and fungi
- "wild" resources (species and other elements)

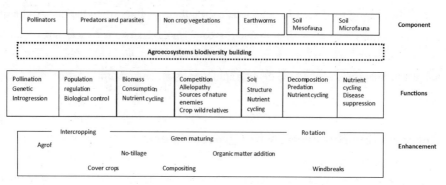

Fig. 10.1　The integration of resources, components and functions for multiple-use farm systems (Altieri, 1995)

10.3.2　*Regenerating the Soil Biodiversity*

　　High soil quality plays a crucial role in agroecosystem stability; however, some agricultural practices make it difficult to maintain an ecologically balanced and productive soil environment. Frequent tillage and chemical overuse reduce soil biodiversity and increase the dependence on a few key crops.

　　As Chap. 3 has discussed in more detail, soil biodiversity regulates a host of activities in the soil ecosystem, which directly affect the fertility and ability of the soil to support plant and animal growth. Rich soil biomass may contain thousands of organisms, fungi and bacteria for decomposition and nutrient cycling. Soil microbes influence plant nutrient availability and increase soil tolerance to disturbances. Table 10.6 summarizes key influences of soil biota on soil processes. Increasing topsoil and soil organic matter will improve biological processes without human input. Managing soil quality will stabilize crop production and nutrient cycling in an agroecosystem and minimize risk of disease and losses on the farm.

Table 10.6　Influences of soil biota on soil processes in ecosystems (Altieri, 1999)

Types of soil biota	Functions in nutrient cycling	Functions in maintaining soil structure
Microflora (fungi, bacteria, *actinomycetes*)	Catabolize organic matter; mineralize and immobilize nutrients	Produce organic compounds that bind aggregates; hyphae entangle particles onto aggregates
Microfauna (*Acarina, Collembola*)	Regulate bacterial and fungal populations; alter nutrient turnover	May affect aggregate structure through interactions with microflora
Mesofauna (*Acarina, Collembola, enchytraeids*)	Regulate fungal and microfaunal populations; alter nutrient turnover; fragment plant residues	Produce faecal pellets; create biopores; promote humification
Macrofauna (isopods, centipedes, millipedes, earthworms, etc.)	Fragment plant residues; stimulate microbial activity	Mix organic and mineral particles; redistribute organic matter and microorganisms; create biopores; promote humification; produce faecal pellets

10.3.3   Increasing the Landscape Biodiversity

Since millions of microbes exist on Earth and perform extremely important functions, many studies focusing on biodiversity regeneration in agroecosystems have been conducted at the field level, rarely considering any larger scale. However, large-scale fragmented landscapes produced by crop monocultures have significantly affected agroecosystem biodiversity and reintroducing a mosaic structure can lead to the creation of multiple habitats for shelter, feeding and reproduction (Altieri, 1999). Some options for regenerating biodiversity at the landscape level include diverse vegetation field margins, head rows, wetlands, woodlots, fence rows and farmyards. Corridors can modify microclimates and air currents, influence nutrient, water and material flows and interrupt disease dispersion. Landscape level biodiversity helps stabilize these processes in the agroecosystem for greater control over the system and minimizing human dependence on technology, labour and other resource rich materials.

10.3.4   Rebuilding Field Biodiversity in Agroecosystems

Restoring biodiversity to agroecosystems can be executed in numerous ways. Identifying the desirable type of biodiversity to carry out specific, intended ecological services will help one determine the best management practices. Well planned strategies can enhance desired functional components of biodiversity in the agroecosystem and implement synergisms through polycultures, agroforestry systems and crop–livestock mixture arrangements (Atlieri, 1999). Fig.10.2 demonstrates different agricultural designs and practices that can enhance functional diversity or negatively affect it. This diagram is most helpful when used in conjunction with a specific agroecosystem goal for enhancing or regenerating biodiversity.

Evidence suggests some farm-management practices addressed in Fig. 10.2 have already successfully made space for wildlife in agroecosystems. These practices include: ①establishing conservation-tillage systems, ②minimizing herbicide application, ③leaving uncultivated strips within crop fields as habitat for weedy relatives of crop plants, ④windbreaks, ⑤border plantings or live fences between plots or paddocks or between farms, ⑥irrigation bunds, ⑦vegetative barriers to soil and water movement within crop fields (Alkorta et al., 2003).

10.3.4.1   Using Multi-Cropping to Restore the Landscape Biodiversity

The use of multi-cropping can turn a once heavily eroded hill valley in Southern China into restored land focusing on watershed management and agricultural production (Fig. 10.3). This photograph depicts interlaced crop species forming forested land, orchards and herbaceous plants. Animal rearing takes place in the valley providing manure for the plants. Through indigenous knowledge, these plants provide medicines and a religious sanctuary for the local people, in addition to food and fibre.

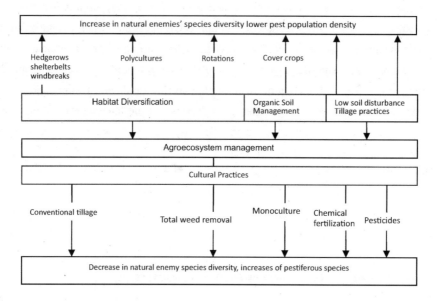

Fig. 10.2   The effects of agroecosystem management and associated cultural practices on the biodiversity of natural enemies and the abundance of insect pests (Altieri, 1999)

Fig. 10.3   An agro-landscape in a hill valley in Southern China (Luo, 2007, personal communication)

Fig. 10.4 demonstrates another field conversion strategy in the Pearl River Delta or Guangdong, China. "Sangji Yutang" refers to the system of mulberry fields and fishponds, where paddy fields are converted to fishponds and surrounded by mulberry plants, sugarcane, grasses or miscellaneous vegetables on elevated land. Typical land use relied on mulberry leaves feeding silkworms to produce enormous quantities of silk for export to Europe and the USA. Excess mulberry fed the pond fish or fertilized the mud below. This well-established field conversion system distinguished the delta regions in south-eastern China, allowing this traditional, family oriented, labour intensive, value-added industrial processing system to survive the Depression, several wars and political upheavals and through the post-Mao years (Hsieh, 2000).

Fig. 10.4   "Sangji Yutang" landscape in the Pearl River Delta or less commonly known as Guangdong, China (Luo, 2007, personal communication)

This system has evolved over 1000 years

10.3.4.2   Managing the Field

Using diversifying methods, such as intercropping and agroforestry, to mimic naturally ecological processes while optimizing natural inputs of sunlight, soil nutrients and rainfall will help to create a sustainably manageable, complex agroecosystem. Assembling a diverse array of functional components in an agroecosystem will initiate synergisms to naturally activate and substantiate ecosystem services. Using different combinations of crops, trees and animals will activate soil biological processes, nutrient recycling and immunity reinforcement. Strategies to restore agrobiodiversity should exhibit the following features (Altieri, 2005):

• Crop rotations: to provide temporal diversity, crop nutrients and to break pest, disease and weed life cycles.

- Polycultures: to introduce competition or complementation to enhance yields.
- Agroforestry systems: to enhance complementary relations between functional components for multiple uses within the agroecosystem.
- Cover crops: to improve soil fertility, biological control of pests and to modify orchard microclimate.
- Crop/livestock mixtures: to aid in the achievement of high biomass output and optimal recycling.

Fig.10.5 demonstrates a rice-fish paddy ecosystem in China. In a traditional paddy field, fish eat weeds and insects reducing the need for pesticide use around the rice. This also diminishes GHG emissions from the paddy soil.

Fig. 10.5　A rice-fish paddy agroecosystem in China (Luo, 2007, personal communication)

10.3.4.3　Managing Soil

Using an ecosystem approach to the application of production techniques and land use practices can enhance soil biodiversity management. According to the FAO, the following techniques can enhance soil biodiversity functions in the production of basic grains. To broaden agroecosystem resilience and improve yields, use crop rotations, leguminous cover crops, improved local seed varieties and diversified crop associations. To reduce disturbances of soil structure and biota, use low-impact tillage methods. To produce organic fertilizers, use stubble, harvest residues, livestock manure and green manure. Finally, to main soil structure and moisture content, use conservation measures.

The FAO also describes the following land management practices that favour plant and animal diversity associated with soil biological activity. The methods include: mosaics of various crops and land uses; capturing and conserving rainwater for plants, animals and people; incorporating backyard animals using their manure for soil organic matter in home gardens and restoring agricultural biodiversity by planting native crops, medicinal plants and tree species.

10.3.4.4　Managing Crops

To reverse genetic agrobiodiversity loss from monocultures, multi-cropping such as intercropping can provide field integrated pest management and increased disease resistance in crop production in addition to improved genetic diversity. An experiment comparing monocultures with intercropping 24 traditional and hybrid rice varieties in the Yunnan Province, China, resulted in a significantly more resistant rice field to epidemic disease when using mixed planting, eliminating the need for pesticide use in these plots (Zhu et al., 2000). Fig.10.6 shows this ecologically sound approach for disease control and decreased pesticide use.

Fig. 10.6　24 varieties of rice intercropping in Yunnan Province, which decreased pesticide use
(Zhu et al., 2000)

Part IV

Application of Agroecosystem Concepts

Chapter 11
Domestication in Agricultural Systems

C. D. Caldwell

"The problem is that only a tiny minority of wild plants and animals lend themselves to domestication, and those few are concentrated in about half a dozen parts of the world."
—Jared Diamond

Abstract Human society is dependent on a very few plants and animals. This chapter deals with domestication by humans of both plants and animals. The differences in adaptations for wild species vs adaptations for domesticated species is related to the separate goals of survival vs production for human goals. The advent of domestication of crops and animals were society-changing inventions from a nomadic lifestyle to a sedentary, more stable one, resulting in the first human population explosion. Animal domestication has ethical, environmental and health aspects to consider and the student is challenged with the question of domestication vs exploitation.

Learning Objectives
After this studying this topic, students should be able to answer the following questions:

1. What difficulties did hunter–gatherers face?
2. What is domestication?
3. Differentiate between adaptations of wild species and domesticated and explain the reasons for the differences, using at least one plant and one animal example.
4. How did domestication allow for an increase in human population?
5. Name at least five recent examples of domestication (or part-domestication) ?
6. Discuss the reasons for and against the consumption of meat by humans.
7. Is domestication exploitation? What is your opinion?

11.1 Introduction

Human society is dependent on a very few plants and animals. While we know of more than 350,000 species of flowering plants and have identified more than 3000 economically important plants, in terms of human diet, we are dependent on 12 plants and three types of livestock for almost all our food and food products. As was discussed in an earlier chapter, ecosystem stability is closely linked to ecosystem diversity. When a species is highly dependent on very few other species, it makes for instability. For example, if there were a devastating disease that could wipe out maize (corn) across the planet, survival of our species would be put in great jeopardy. In a less dramatic way, diseases attacking banana, cocoa or tea have a significant effect on human culture, livelihood and economics.

This chapter deals with domestication by humans of both plants and animals. Consider for a moment if you were marooned on a desert island and had to find something to eat. If you are hungry, what would you try first? Would you eat a plant you did not recognize, or would you think it better to try eating an unfamiliar animal? Most people would answer that they would first try eating an unfamiliar animal and, in general, that would be the safest choice. Most wild plants have some sort of selfdefence mechanism to discourage herbivory; this may take the form of thorns or prickles but often plants choose to have some sort of chemical defences. These are common plant toxins that can either just give a bitter taste, perhaps as a warning to the herbivore, or maybe very poisonous. Since plants cannot run away from their predators, they must have other ways of survival. Therefore, when we select a prospective plant for domestication for food purposes, it usually involves selection of types with low toxicity and ease of harvest. The earliest plant breeders were our ancestors who lived as hunters and gatherers. Through trial and error, selection and replanting, followed by more selection, our first crop plants were domesticated.

Domestication may be defined as the process of establishing a relationship between people and a plant or an animal in which the plant or animal is no longer "wild" but instead lives in association with people, for at least part of its life cycle, and provides some benefit to people. A defining characteristic of domestication is artificial selection by humans.

11.2　Adaptations for Wild Species Versus Adaptations for Domesticated Species

From a species standpoint, a successful wild plant or animal is one that survives and can reproduce and carry on the genes for the next generation. This goal of survival and reproduction is therefore the driving force for adaptations of a wild species, either plant or animal. By contrast, the overall goal of a domestic plant or animal is to be of some use to humans. Obviously, a domesticated plant or animal needs to have traits that confer survivability. However, those traits are within the overall care of the farmer. In other words, the survival traits are not as strong when there is protection. On the other hand, the adaptations of a domestic plant or animal mean that they will be productive in terms of materials for human consumption or sale.

This can be illustrated by looking at Fig. 11.1 that shows a picture of a bison and that of a dairy cow. The bison can be used as an example of a wild animal and the dairy cow is an example of a domesticated species. Adaptations can be generally divided into three types: structural, metabolic/biochemical and behavioural. Looking at the picture of the bison, one can see immediate structural adaptations for survival. This is a large, strong animal with horns for protection against predators. It has a heavy coat which gives it protection against the cold. From a metabolic standpoint, this animal is a ruminant and is adapted for grazing and browsing. The rumen in it produces heat during the digestive process that keeps the animal warm in the cold winters in which it inhabits. Behaviourally, the bison is known to be very protective of its young. The adults will form a circle around the young of the species to protect against predators such as wolves. It has developed aggressive behaviour to protect itself and its family.

Fig. 11.1   Bison vs dairy cow

By contrast, consider the dairy cow. In terms of survival, this is a species that is well protected by the farmer. Housing and protection from predators are provided; this means that structurally, the adaptations are more towards production of milk and not for individual survival. Hornless cattle have been bred and those that do produce horns are usually dehorned in order to protect the farmer. Dairy cattle have been selected for ability to produce calves with little birthing problem, good feet resistant to being in barns and not in soft fields and ability to produce lots of milk. From a metabolic standpoint, the dairy cow is also a ruminant with the ability to digest grasses. From a behavioural standpoint, aggression and aggressive behaviour have been bred out of the species. Dairy cattle in general are very passive; any cow that shows aggression will soon be culled from the herd.

A similar analysis can be made in comparing wild and domesticated plants. Again, a wild plant is successful if it survives and reproduces, while a domesticated plant is one that provides a significant economic benefit to the farmer. Fig. 11.2 compares a wild oat to a domesticated oat. Wild oats have characteristics such as shattering; i.e., as soon as the seeds are mature, they fall from the plant. Domesticated oats do not have this characteristic since we want to have the seed stay on the plant until the farmer is ready to harvest. Wild oats have staggered dormancy; i.e., they will germinate over a period of several years once they are mature. Domesticated have a synchronized dormancy which is very short so that the farmer is able to reseed and knows that they will get good germination. Wild oat seeds are quite small compared to their domesticated cousins.

These are just examples of the types of differences between wild and domesticated types that illustrate the sort of selection pressure that is happened to move them from my wild state to a useful, domesticated state.

Fig. 11.2　Wild oats (A) vs domestic oats (B) (photo credit: C. D. Caldwell)

11.3　How Did Domestication Allow for an Increase in Human Population?

Consider for a moment the lifestyle of our ancestor hunter-gatherers. There were not any domesticated animals 10,000 years ago nor probably any domestic crops or field or garden crops. The human population of the Earth was much smaller than it is now, perhaps less than one million people total on the earth. What would people have been eating at that time? They would gather wild fruits, wild berries and vegetables growing in nature; hunt wild terrestrial animals; fish, if they were close to the water and they would probably raid birds' nests for eggs. What might be the difficulty of making a living in that way?

By necessity their lifestyle was nomadic. Relying on hunting and gathering in one place meant depletion of resources, so it was necessary for groups of huntergatherers to move from place to place perhaps circling around to return to familiar hunting and gathering areas when there was recovery from their activities. There would be times of great plenty when there were lots to eat in the wild but there would also be times during the year when it was difficult to hunt and gather, particularly in colder climates. From a family and community standpoint, this means that families were relatively small, since moving with very young children would be a challenge. Hunting wild animals also is a dangerous undertaking, especially for larger prey. Domestication of animals and plants allowed families and family groups to build more permanent living areas. This led to permanent villages with secure areas and less fluctuation in food availability. As one would expect, this allowed for larger families and more children to survive. The advent of domestication and sedentary agriculture, therefore, provided the opportunity for food security and an increase in the human population.

11.4   Domestication and Evolution of Agriculture

It would be natural to attempt to develop an alternative strategy. In a hunter-gatherer society, probably hunger would have been the main limiting factor to numbers. People would often have been close to starvation and death rates were high, thus limiting the size of a human population. In response to those pressures, the domestication of crops and animals were society-changing inventions. Almost certainly, the first plants to be domesticated would have been from the cereal family which includes rice, wheat, barley and rye. These cereals are grasses that have been selected over thousands of years from wild grasses in Asia and Central America. These domesticated grasses are now our most important sources of food. They were the easiest to domesticate because the seeds are annual in growth habit, compact, high in nutrition, lightweight and easy to transport. They could be easily carried with them in their nomadic times and replanted over and over. This provided an opportunity for the earliest selection pressure for large seeds, good taste and good nutrition. It was not until much later that perennial crops would be domesticated when communities were able to become more stable and sedentary to allow for such a domestication.

As for domesticated animals, probably the dog was the first domesticated animal anywhere in the world (Table 11.1). The first dogs were probably rescued young animals that had been orphaned. They evolved into companions and protected animals for families or groups of people. Domestication of food animals and animals for transportation followed over the next thousands of years. Some farm animals have only recently been persuaded to adapt to human control. One example of recent domestication for farming is the eland. The eland is the largest species of antelope; in Kenya and some other East African countries, there are now large eland ranches where it was captured from the wild and persuaded to at least put up with being controlled by humans.

In New Zealand, there are many recently domesticated Red Deer farms or ranches. Two other examples of recent domestication took place in Prince Edward Island of Canada around the year 1900; this is the mink and fox. These two animals can be considered only loosely domesticated; human handling of them can be very dangerous.

Table 11.1　Approximate times for domestication of animals

Domesticated animal	Years since first domestication
Dogs	12,000
Sheep and goats	11,000
Cattle and pigs	9000
Oxen	6000
Cats and horses	5000
Silk moths	5000
Camels, elephants and poultry	4000

11.5　Is Animal Production Good for Humans and the Planet?

Domestication of animals and the incorporation of animal husbandry as an overall part of agricultural systems has a long history. While methods of producing animals and animal products vary across the globe, animals as part of agriculture have been a key component across cultures and geography. Usually, animal production includes livestock rearing, poultry raising and aquaculture. The collective term livestock includes any number of animals raised for various products, including food, fibre or a plethora of by-products. In some societies, livestock are also still used for labour. Red meat types include cattle, water buffalo, sheep, goats and pigs; poultry is the category of domesticated birds including chicken, duck, emu, goose and turkey. These are raised both for meat and for eggs.

There are ongoing discussions concerning the reasons for and against the consumption of meat, milk and eggs by humans. In general, there are four areas of discussion: nutrition, economics, environment and ethics. In terms of nutrition, the consumption of animal products provides high-quality good match of amino acids for production of proteins in human bodies. There are vitamins including B12 and minerals including iron in red meats and zinc. Fish, such as sardines and salmon, are a good source of omega-3 fatty acids and there is conjugated linoleic acid in ruminant meat and milk. Animal products, however, provide no fibre, vitamin C calcium or the phytochemicals that are so necessary for our diets.

The economic environment for meat and edible animal products seems to be in flux and it is varying across the globe. In general, demand for meat products across the world has increased as countries have become more prosperous. However, as people become more aware of health and environmental issues, demand is decreasing for meat products, which are considered either less healthy or detrimental to the environment. Demand, price and cost of production are the controlling factors.

Chap.8 has discussed the environmental impacts of animal agriculture on the environment. The production of animals, especially ruminants, greatly increases carbon footprint of our food production systems. However, it is important to consider the positive side of animal production as well. Having animals that can graze and browse plants that humans cannot eat is a way of harvesting energy and nutrients unavailable to us otherwise. Therefore, the use of marginal pasture and range area can be of benefit to overall cropping systems. Within mixed farming systems, use of animals for nutrient cycling (i.e., manure management) is a crucial way of closing nutrient cycles and preventing nutrient loss from systems. As has been mentioned earlier in the text, feeding animals does compete with humans for energy and nutrients when we feed materials that could otherwise be eaten by people. The higher people eat on the food chain, the greater is the energy loss in the system. Animal systems are also significant users of water and producers of greenhouse gases.

From an ethical standpoint, animal agriculture has several challenges, which are discussed in Chap. 11 concerning animal welfare and is considered below in the question of whether domestication is exploitation.

11.6   Is Domestication Exploitation?

The relationship between domestic animals and people can be explained in terms of the ecological term "symbiosis". Some would argue that domestication results in a mutualistic relationship and others would say it is exploitation. The former is defined as a symbiotic relationship in which both participants benefit (mutualism); the latter is defined as a relationship between two organisms in which one organism benefits, while the other is harmed in some way.

Consider cattle for an example; people eat cattle and take milk from them and use the skins for jackets or shoes or purses. Meanwhile, people do give them something in return by protecting them from predators, giving them food, growing crops for them and giving them shelter by building them barns or windbreaks. Socially, many farmers form a close relationship with their animals. For example, the relationship between people and horses or dogs may be a bit unequal but there is give and take on both sides. Dogs and humans get on particularly well; for some reason, these 2 species suit each other, and they can develop close bonds. The dog is a very versatile animal; this ranges from avalanche dogs to drug-sniffing dogs at airports or sheepdogs or seeing eye dogs. Dogs can do a lot of things very well and people make use of their superior abilities, superior to us in many of these situations. Most dogs seem "satisfied" with the trade-off in terms of their care and the relationship with their owner.

Some domestic animals are wild animals that have only a casual contact with humans. Sometimes people who enjoy sport fishing stock a lake with fish. People can perhaps claim that they are domesticated, and they are in a close relationship than the wild fish, albeit not a relationship to the liking of the fish. Those fish are not domestic but stepping away from being wild fish and towards being domestic species. Consider another example of eiderdown; eiderdown comes from the eider duck that lives in Iceland and a number of other places. This bird eats shrimp and small fish in the sea and it nests on the shore. The duck will line its nest with down, fluffy feathers plucked from its own body and these outer feathers are one of the lightest and best heat insulators known. Eiderdown has a very high heat insulation value. People use eiderdown from these ducks in products like parkas, sleeping bags and comforters. In Iceland, people go and steal some of the down from the birds' nests and the birds then replace it. If you get too greedy and steal too much, then the birds move away. People are exploiting the ducks but in return we are offering them something; in Iceland, humans mount a 24-hour guard on the nesting grounds complete with rifles to keep away other predators, mainly foxes. It is an arrangement with some benefits in both directions. It is not pure exploitation but a form of mutualism.

Another instance of animals that are domestic but only barely so is mussels (Fig. 11.3). There are quite a few mussel farms around the world and many of them are family businesses run with a small boat. The mussel is a shellfish, a type of filter feeder mollusc. It draws a current of seawater through it and filters out tiny particles of plankton which it eats. Mussel farmers hang plastic socks in the sea, a net that stays down the water most the time. The baby mussels attach to the net and they grow on the surface provided by the mussel farmer. Eighteen months later they have grown into big mussels and are ready for harvest. The farmer hoists the nets and harvests the mussels.

Fig. 11.3　Mussels on socks at harvest (photo credit: N. Firth)

Oysters are also filter-feeding shellfish; oyster farmers do more for the oysters than the mussel farmers do. Oyster farmers rear the young oysters as well as growing the bigger ones; therefore, it is said that farmed oysters are more domestic than the mussels. Oysters feed on algae and the oyster farmers grow algae to feed the young shellfish. Baby oysters are grown in tanks, with water circulating and as they grow, they are put into Pearl nets, mesh bags. These are put into the sea and then they filter feed naturally. As oysters grow, the farmer eventually moves them to a different container called a lantern neck (Fig. 11.4) that looks a bit like a lantern and again is lowered to the sea and again the animals will filter feed. The oysters reach market size in 3 or 4 years. Obviously, the farmer does a great deal to assist the survival and growth of the oysters but in return does harvest them for human consumption.

Another recently domesticated animal is the pine martin (Fig. 11.5), utilized for its high-quality fur. They are cute but they also have very sharp teeth. They are semiwild animals; you must wear gloves and take a great deal of precautions dealing with them. Getting wild animals like pine martin to breed in captivity is often difficult and is one of the main obstacles to domestication. People may find out what the animals like to eat in order to feed them successfully; however, it is most important to find the way they breed. Otherwise one cannot keep domestic animals on an ongoing basis. Only recently can people manage to get the martin bred indicating that they are adapting to life in captivity. Would the pine martin rather be free in the forest? Almost certainly, the answer is yes.

Many of our domesticated animals are either birds or mammals. Some of them are very highly domesticated and cannot survive by themselves in the wild. For example, a domesticated turkey has been selected for traits that make it unsuitable for survival outside the barn. They do not have the skills and the physical conformation to survive by themselves. However, some domesticated animals can survive in the wild. There are a number of animals that were once domestic but have returned to the wild; we call these feral animals. In many cases, there are islands with resident feral goats. These goats were originally put there by people and now they take care of themselves. This is also true of horses marooned on islands from shipwrecks or the wild chickens of Hawaii.

Examine the list below in which we have indicated several examples where animals are used to the benefit of humans. Consider which of these examples you would deem to be a mutualistic relationship and which are purely exploitation by humans. Keep your ideas in mind as you consider Chap. 11 regarding animal welfare in agriculture.

Consider which of these examples are exploitation in your opinion:

- Raising animals for food and clothing
- Companion animals
- "Recreational" hunting and fishing
- "Food" hunting and fishing
- Seeing-eye dogs; police dogs
- Animals in Rodeos and circuses
- Animals used for testing medicines, cosmetics and household products
- Animals used for medical and other scientific experiments

Fig. 11.4　Lantern neck for oysters (photo credit: N. Firth)

Fig. 11.5　Pine martin (photo credit: N. Firth)

Chapter 12
Animal Welfare: A Good Life for Animals

T. Tennessen and C. D. Caldwell

"The greatness of a nation and its moral progress can be judged by the way its animals are treated."
—*Mahatma Gandhi*

Abstract Animal welfare means concern for the quality of life of animals under our control. There are many issues surrounding the welfare of animals in agricultural systems; those issues usually can be classified within the general context of either housing or production procedures. Because people cannot talk directly with animals, it is difficult to determine if animals have a good life. However, using our somewhat subjective judgment and ethics combined with observation using proven scientific methods, some answers are emerging. Apart from using critical observation, there are also scientific approaches to assess quality of life or animal welfare. The main tools are the study of animal behavior, physiology, and epidemiology. We often use the health status of animals in confinement and production systems as indicators of well-being. These scientific approaches can give us some objective clear data to help us make our management decisions or help animals to cope with confinement better. Another approach commonly used by animal scientists is to assess the animals' environment: the physical environment, the social environment, and the quality of the animal–human interaction.

Learning Objectives

After studying this topic, students should be able to:

1. Describe the range of attitudes from "animal welfare" to "animal rights" as those terms are used in the present-day debate on the use and care of animals in agriculture.
2. Describe the principal animal welfare issues in modern animal agriculture. Illustrate using examples from poultry, swine, beef cattle, fur, and dairy farming.
3. Explain the importance of the physical environment, social environment, and the human–animal relationship to farm animals' quality of life.

12.1 Introduction

In the current debate about animal welfare, there is a spectrum of opinions among those who are concerned about animals. Some consider themselves animal welfare advocates and others take a more radical animal rights position. Animal welfare means concern for the quality of life of animals under our control. A more extreme position is that of animal rights, which asserts that animals are separate nations and should be free to live lives unencumbered by human interference. Animal welfare advocates would allow the raising of animals for food as long as their quality of life is good, and slaughter is humane. Animal rights promoters would not allow raising animals for food and many who hold this view would also be opposed to keeping animals as pets. In discussions of agricultural systems, the concern is usually focused on animal welfare rather than animal rights; i.e., the concern for the quality of life of animals under our control while we raise domesticated animals for food. However, the positions taken by animal rights activists have influenced our culture by making agriculturists examine more closely all animal care.

12.2　Principal Animal Welfare Issues in Modern Animal Agriculture

There are many issues surrounding the welfare of animals in agricultural systems; those issues usually can be classified within the general context of either housing or production procedures.

Examples:

Particular attention recently has been paid to such concrete examples of housing issues as confinement crates for veal calves, small cages for laying hens, gestation stalls for sows, long-term tethering of dairy cows, and the caging of foxes and mink in fur farms. Fig. 12.1 shows a crate for a veal calf. These are usually culled dairy bull calves that are housed in small pens and often fed diets that are liquid and that may be somewhat low in iron, so as to produce a paler meat. The lack of exercise also tends to produce a more tender meat than that of veal calves that are given roughage to eat and opportunity to move freely. Many people consider this type of treatment to be cruel and unnecessary.

Battery cages for laying hens is one of the main sources of criticism from animal welfare groups throughout the world, especially in North America and Europe. These are small cages in which anywhere from 3 to 6 birds are kept for their whole adult lives (approximately 1 year) to lay eggs. In many cases worldwide, the space per bird is about 450 cm^2 per bird. Animal welfare pressure has meant that some countries are responding and changing that regulation; for example, effective 2022, all cages for layer hens in Canada must provide 750 cm^2 per bird. Despite considerable evidence that hens suffer as a direct consequence of battery systems, most countries have not addressed this welfare issue with any significant legislation or regulation.

Fig. 12.2 demonstrates a gestation stall for sows. Gestation refers to the pregnancy period of the sow and during the sow's pregnancy she may be kept in a small stall like this where she is tied or confined by bars. It does not allow her any exercise; it does not allow her to interact with other animals of her kind (pigs are very social animals) and there may be serious discomfort due to the confinement and the tether.

Fig. 12.1   Crate for a veal calf (photo credit: T. Tennessen)

Traditionally, the dairy industry is seen as a model of the classical farm animal outside on pasture; it looks like everyone's idea of the agrarian ideal. However, dairy cow systems are becoming more intensive, although the barns usually are of high quality and the feed is of excellent quality. In some dairy systems, there is long-term tethering of dairy cows and this is a welfare problem. In some dairy production systems, the cows are not put on pasture; in North America, this is called a zerograzing system and if the barn is of a Tie-stall design, then the dairy cow may be tethered and kept on a rather short tether for many weeks without having any chance to move freely.

Our last example is caging of foxes and mink for the purpose of production of fur. Keeping animals for fur production is often viewed quite differently than keeping animals for food production. Food is a necessity; furs are not. Furs can be replaced by synthetics; it can therefore be argued that this is a luxury that we should end. To keep foxes and mink in a healthy environment without parasites, a common solution has been to keep them in cages off the ground separating the foxes and mink from their feces. This helps to control parasites but also gives a very simplified, non-enriched environment without much stimulation for the animals. It does not allow interaction with others of their kind.

So these few examples are some of the main housing concerns when it comes to farm animal welfare: housing of laying hens, sows, and veal calves; long-term tethering of dairy cows; and the keeping of foxes and mink on fur farms.

Many other animal welfare issues in modern animal agriculture have to do with procedures.

Cattle are subjected to castration, branding, dehorning, and ear notching. Pigs are subjected to castration, teeth clipping, mixing with other animals, especially at a young age and early weaning. The mixing and the early weaning are very difficult for pigs since they are such social animals. Poultry are subjected to beak trimming, disposal of day-old chicks (males) and spent hens (hens that have completed their year of egg laying), and other procedures that may cause pain and discomfort for the animals. They are most often done without anesthetics and contribute to the sometimes-widespread criticism for the way animals are reared for food production. However, in general, the consumer's demand for cheap food is often at odds with concerns over animal welfare.

Fig. 12.2   Gestation stall for a sow (photo credit: T. Tennessen)

12.3   Animal Welfare: Observations and Research

Because people cannot talk directly with animals, it is difficult to determine if animals have a good life. However, using our somewhat subjective judgment and ethics combined with observation using proven scientific methods, some answers are emerging. In many cases, our intuition is a reasonable guide. For example, Fig. 12.3 shows cattle grazing on a broad pasture by the ocean. These cattle are in an environment close to the natural environment, an environment to which they are well adapted. There is complexity to the environment; they are able to interact with others of their kind and express natural behaviors.

By comparison, cows overwintering in snowy woods (Fig. 12.4) may be judged by some to be in a cruel situation because it appears that the cow is in an inhospitable environment with so much snow; however, ruminants are well adapted to the cold. The cow has shelter in the forest and can eat snow as a source of water the way wild ruminants do. By way of contrast, cattle in a feedlot (Fig. 12.5) although they have apparently lots of food, may show the problems of lack of environmental complexity, forced crowding, mixing of cohorts, and aggressive interactions. Environments like these begin to pose some problems for animal welfare. From a production standpoint, the feedlot is an efficient system for growing cattle at an optimum rate of weight gain for maximum profit whereas housing cattle in a forest situation or on large pasture can be less rewarding financially.

Fig. 12.3　Cattle grazing near the Bay of Fundy in Nova Scotia Canada (photo credit: T. Tennessen)

Fig. 12.4　Cattle overwintering in a snowy wooded area (photo credit: T. Tennessen)

In Fig. 12.6, the dairy cow in a tie-stall looks content; she has comfortable bedding, enough food and water and is kept healthy through regular care. One does observe that she is on a short tether and the question is can she adopt comfortable resting positions on the short tether? Is this overall a good situation that allows for the animal to express its natural behaviors? How often the cow is released and allowed to mingle with other cows is an important consideration for animal welfare.

Fig. 12.5　Cattle in a feedlot in Western Canada (photo credit: T. Tennessen)

Fig. 12.6　Dairy cow in a tie stall (photo credit: T. Tennessen)

12.4　Scientific Research

Apart from using critical observation, there are also scientific approaches to assess quality of life or animal welfare (Fraser, 2009). The main tools are the study of animal behavior, physiology, and epidemiology. We often use the health status of animals in confinement and production systems as indicators of well-being. These scientific approaches can give us some objective clear data to help us make our management decisions or help animals to cope with confinement better. Another approach commonly used by animal scientists is to assess the animals' environment: the physical environment, the social environment, and the quality of the animal–human interaction. We shall use the example of sows to demonstrate these approaches.

12.4.1   Physical Environment

How much space do animals need? Sows during their pregnancy are often housed and tethered. The gestation period of the sow is about 115 days and for much of this time, the sow may be confined with no opportunity to exercise or to exhibit natural behaviors; in fact, sows often show repetitive stereotypical behavior while in confinement, such as chomping on bars. Normally in the wild, a sow would try to build a nest during her gestation. Sows in confinement will sometimes go through some of the movements but without access to nesting materials, are not able to fulfill that behavioral drive.

The next consideration for the mother pig is how much space and what structures are needed for a farrowing sow. The answer has to consider the comfort of the sows and the safety for piglets, i.e., they have to be protected from being crushed by the sow when she lies down. Pigs are an animal where the adults are very big, and the newborns are very many and very small. It is very easy for piglets to be crushed accidentally by their mother. The domestic sow is so big that when she lies down she does so very quickly and may fall on top of piglets, suffocating them in the process. Farrowing crates are constructed to protect the piglets from their mother. So, it is probably a fairly good environment for the piglets; however, the sow can only do 2 things, stand up and lie down. She cannot interact with them beyond nursing; she cannot turn around and she is limited in how comfortable she can make herself for the duration of the nursing, which is usually about 3 weeks.

Some modified farrowing crates allow the sow to get up and lie down and also allow her to turn around and make herself more comfortable. The farrowing crate is a bit broader yet it also gives protection to the piglets from accidental crushing. The improved crate shown in Fig. 12.7 is a Norwegian farrowing pen. These are commonly used in the pig production industry in Norway and provide more freedom of movement. The sow can explore her surroundings a little bit, she can turn around, lie down, stand up, and there is also a small area in the corner underneath the heat lamp where the piglets can rest and get warm and can be protected from being accidentally crushed by the mother.

12.4.2   Environmental Enrichment

Besides the size, quality of space is also important for farm animals. The crate in Fig. 12.8 includes some straw bedding and, while there is no area where the piglets are totally away from the mother, there is enough complexity here. Enough small corners, combined with straw make this a good system. Given sufficient straw, even if the sow accidentally lies down on top of the piglets, there is usually enough cushioning that she can get up again and the piglet may not be injured.

Fig. 12.7　A Norwegian farrowing pen (photo credit: T. Tennessen)

Fig. 12.8　A farrowing area that includes some straw bedding for complexity (photo credit: T. Tennessen)

　　In short, the quality of space depends much on environmental enrichment, which is usually described as "bringing nature into captivity". Environmental enrichment should also be functional. Fig. 12.9 demonstrates a sow with her head just poking out from the straw which is a good material for regulating body heat, so that they can dig themselves down into it if the weather is cold. They can also eat some of it, they can root in it, and rooting with their snout is such a common behavior of pigs in the wild that to entirely prevent pigs from doing that might be unreasonable. Pigs seem to greatly appreciate having some opportunity for rooting behavior. If provided with straw, a sow will make a nest before farrowing. If a sow that is about to farrow is housed in a barn that has a lot of straw, so she can carry that straw around and will actually build a nest or cushion area where she will give birth.

Fig. 12.9　Sow demonstrating nesting behavior in deep straw bedding (photo credit: T. Tennessen)

Fig. 12.10 shows a farrowing system with both ample physical space and environmental enrichment. The sows are in groups with the piglets, so in this group there may be six sows each with about ten piglets, so that there is a group of six adults and about 60 young pigs. Pigs are very much a communal animal and this really helps build a high quality of life. They interact with others of their kind and they have a complex environment with which they can also interact.

Fig. 12.10　Farrowing system that shows both physical space and environmental enrichment

(photo credit: T. Tennessen)

In nature, sows live in groups. Free-ranging and feral pigs in various parts of the world organize themselves into groups of sows, maybe 10 or a dozen or so with all their piglets. The production system shown in Fig. 12.10 demonstrates such a production system. The environment is socially complex and physically complex. The sows do not fight and they do not hurt each other's piglets. One can add even more environmental complexity by bringing the animals outdoors and giving them more space to move around. There are problems with housing pigs outdoors in some very hot or cold climates and ease of handling by the farmer is reduced as freedom of movement of the pigs is increased.

12.4.3　The Animal–Human Relationship

How humans interact with animals is important; the nature of their relationship will mean that the animals can be at ease, content, calm in the presence of humans or be nervous and frightened, and stressed.

The nature of the relationship can affect the productivity and the welfare of farm animals. Such relationships are formed by way of the interactions between humans and animals and these interactions can be positive, neutral, or negative.

An experiment by David Fraser at the University of British Columbia in 2009 illustrates what happened when sows were given either pleasant handling, in which the handler would walk and squat in the pen and allow sows to approach and interact; or given minimal handling, with no exposure to humans except during feeding and weighing; or were given unpleasant handling where the handler stood in the pen and delivered a brief shock if the sow approached. The results of this experiment showed that sows that were handled in a pleasant way were more likely to become pregnant when bred than the other two treatments. Those sows handled in an unpleasant way were perhaps stressed and perhaps the physiology of stress which may counteract the physiology of growth and reproduction, had an effect.

Furthermore, pigs showing a high level of fear of humans had up to an 11.3% reduction in growth rate showing again that the physiology of stress and the physiology of growth are at odds and contradictory to each other in many ways. This should make us rethink how we handle animals and the kind of relationship we have with our livestock both for ethical and economic reasons.

In summary as managers of animals, farmers control many factors that are very important to the welfare of their livestock and good welfare practices can result in good financial results.

12.5　International Approaches to Implementation of Animal Welfare and Their Impacts

While governments have been slow to respond to issues of animal welfare, increasingly consumers have become concerned. Particularly in developed countries, food safety and quality are being linked to animal welfare indicators. Consumers are also expressing dissatisfaction with animal production systems on an ethical basis. This has created economic incentives for businesses to meet higher animal welfare standards and it has put pressure on governments to establish legislation linked to better animal welfare.

These concerns are expressed in the Five Freedoms for animals under our care (UFAW, 1999): (Universities Federation for Animal Care, United Kingdom)

1. Freedom from hunger and thirst—By ready access to freshwater and a diet designed to maintain full health and vigor.
2. Freedom from discomfort—By the provision of an appropriate environment including shelter and a comfortable resting area.
3. Freedom from pain, injury, or disease—By prevention or through rapid diagnosis and treatment.
4. Freedom to express normal behavior—By the provision of sufficient space, proper facilities, and company of the animal's own kind.
5. Freedom from fear and distress—By the assurance of conditions that avoid mental suffering.

Legislation and Regulations

There are many organizations throughout the world concerned with animal welfare; however, the implementation and enforcement of legislation and regulation for the well-being of animals have fallen to both national and local governments. There is a great deal of variation across the world in terms of animal welfare enforcement.

Globally, the lead organization with responsibility for animal welfare is the World Organization for Animal Health (OIE). Its membership includes 182 countries and it should be noted that its recognition in terms of animal welfare shows how the member countries perceive the close relationship between animal health and animal welfare.

At the national level, countries have different strategies for dealing with animal welfare issues. Some will use national legislation while others will disperse responsibility into individual provinces or states. For example, Canada is a nation in which the legislation, both at the national and at the provincial level, is good; however, the enforcement is not always effective, even in such a developed country. Having diverse legislation and enforcement across the different provinces and territories does not always give consistent protection of animals, including farm animals. One positive note from the Canadian example is that the willful causing of suffering to animals, either through neglect pain or injury is considered to be a criminal act. This law has been enforced quite a number of times to the benefit of farmed animals.

More recently in 2016, China produced its first draft animal husbandry and slaughtering standards code developed under the Chinese Veterinary Medical Association (CVMA). It is just now (2020) working on implementation with 30 domestic livestock breeding and slaughtering enterprises to draft the standards.

There is much more to be done worldwide to enhance the care and well-being of animals produced for food and recreational purposes.

Chapter 13
Forage-Based Production Systems

N. Mclean

"The agriculture we seek will act like an ecosystem, feature material recycling and run on the contemporary sunlight of our star."
—*Wes Jackson*

Abstract Forage crops are grown to feed animals in the form of pasture, hay, and silage. They cover more land than any other type of agricultural crop. Forages are important components of healthy agroecosystems by providing benefits beyond feed for animals. They play a significant role in maintaining healthy, sustainable agroecosystems for long-term stability both economically and environmentally.

Learning Objectives

At the end of studying this topic, students should be able to:

1. Define, with examples, forage crops.
2. List and describe five ways in which forage-based cropping systems can positively impact the environment.
3. Differentiate between forage grasses and forage legumes in terms of their roles and relative economic and environmental advantages in a rotation.
4. Compare three different methods of harvesting/storing forage crops.

13.1   Introduction

Forages are those crops, apart from grains, that are grown to be fed to livestock. Some livestock, including ruminants, camelids, and horses, may obtain all or most of their diet from forage crops while monogastric livestock, such as poultry and swine, can use forage for a portion of their ration but they require additional materials, such as grains, for starch, and additional protein. Ruminants, which include cattle, sheep, goats, and deer, have a four-part stomach and the largest part is the rumen which contains microbes that break down fiber and release volatile fatty acids for use by the animals. The microbes also can produce protein from nonprotein nitrogen sources and this microbial protein is available to the host animal when microbial cells die. Camelids, which include alpacas, llamas, and camels, have a three-part stomach which also makes use of microbes to break down fiber and release volatile fatty acids and provide additional protein. Horses are less efficient at digesting fiber because the microbial breakdown of fiber occurs in the caecum, which is after the stomach, and passage of food through the digestive tract is much faster in horses than ruminants or camelids due to limited volume.

Ruminants, camelids, and horses have evolved on forage diets. While milk and meat are sometimes viewed as less than ideal components of human diets, ruminants fed solely, or primarily forage crops (i.e., little or no grain), provide milk (reviewed by Elergsma, 2015), and meat (reviewed by Daley et al., 2010) with healthy fatty acid profiles including significantly higher proportions of omega-3 fatty acids and CLA (conjugated linoleic acid, also known as rumenic acid), which have shown a variety of health benefits (reviewed by Gómez-Cortés et al., 2018).

Forage crops are primarily members of two plant families: Poaceae (grasses) and Fabaceae (legumes). Species in other plant families are also grown as forage crops, including some members of the Brassicaceae; however, this is on a much smaller scale. There is a wide diversity within both Poaceae and Fabaceae, including a range of both annual and perennial species. In addition to their importance in feeding livestock, these three plant families are also the most important plant families for providing food and other products for humans. Worldwide, alfalfa (*Medicago* spp.) is the most widely grown forage legume, while other commonly cultivated herbaceous forage legumes include species within the genera *Trifolium*, *Vicia*, and *Lotus* (Capstaff and Miller, 2018). Some tree legumes, including *Acacia* species and *Leuconema leucocephala* are used as forage crops in the tropics (Capstaff and Miller, 2018) while temperate forage production relies mainly on herbaceous legumes. The most common grasses in temperate climates belong to the genera *Agrostis*, *Festuca*, *Lolium*, and *Dactylis* and the most common grasses in the tropics are *Pennisetum purpurea* and species within the genera *Brachiaria* and *Panicum* (Capstaff and Miller, 2018).

The majority of forage cropland is planted to perennial grasses and legumes; however, annual forage crops are also common. Corn/maize (*Zea mays* L.) is an annual grass with C_4 photosynthesis which is managed as a silage crop, or, less commonly, for pasture. Other C_4 annual grasses that are grown as forage are pearl millet [*Pennisetum glaucum* (L.) R. Br.], sorghum [*Sorghum bicolor* (L.) Moench], sudangrass [*Sorghum bicolor* (L.) Moench], and sorghum–sudangrass hybrids. C_4 plants are adapted to relatively hot and dry conditions and produce higher yields than C_3 plants under those conditions. Annual forage crops, including forage brassicas, can be used to extend the grazing season in areas where perennial forages go dormant due to freezing temperatures in the autumn.

13.2　Forages in Production Systems

Natural grasslands share the characteristic of being too dry to support forests and too wet for deserts. Forty percent of the Earth's landmass, excluding Greenland and Antarctica, are natural grasslands (White et al., 2000). Natural grasslands exist on all continents except Antarctica, and native plants growing there can be harvested to feed animals, either by grazing or by mechanical harvesting equipment. Natural grasslands include the North American prairie, the South American pampas, the Asian steppes, the African savannah, the South African veldt, and the Australian plains. Natural grasslands are still in use for native forage crop production; however, some natural grasslands have also been converted into use for cereal and horticultural crops, as well as seeding to nonnative forage species.

Forage crops are also grown on land where the natural vegetation is forest. Land is cleared and forage crops (as well as other crops) are planted with species that are not native to the area. Yields are often higher than from natural grasslands due to increased soil moisture.

Forage crops are often grown on land that is not suitable for other crops due to limitations including slope, stoniness, poor drainage, and low pH. Some cool-season perennial forage crops are well-adapted to these conditions and can be grown to feed animals. In this way, forage crops can capture the sun's energy in places where field crops cannot and indirectly provide products for humans such as meat, milk, fiber, leather, work (oxen and horses), and recreation.

13.3　Beneficial Effects of Forages

Forages have several beneficial impacts on the environment compared to cereal and horticultural crops. Environmental benefits are greater for perennial stands than for annual stands because the soil is covered with vegetation throughout the year, reducing soil erosion, and there is no need to reseed yearly, which reduces greenhouse gas emissions related to equipment use. Beneficial impacts include carbon sequestration in organic matter and release of oxygen; protection of the soil from wind and water erosion; protection of waterways from sediment and chemicals; improved soil in terms of soil organic matter level, improved soil structure and improved soil fertility; and provision of food and shelter for wildlife.

Natural ecosystems are self-sustaining due to ①energy capture through photosynthesis, ②energy flow through food chains, and ③nutrient recycling via decomposition and mineralization. Forage-based cropping systems often mimic natural ecosystems in this way, but forage-based ecosystems become unsustainable when nutrients are added in excess of what is removed; e.g., when particular mineral fertilizers build up in the soil or when the level of nutrients removed in the crop exceeds the level of what is returned to the soil through manure, compost, fertilizer, or other soil amendments.

Grasses and legumes are among the top five plant families in terms of species diversity. Legume species include herbs, shrubs, trees, and vines. Forage species within this family have the advantage of a symbiotic relationship with bacteria in root nodules. The bacteria can capture nitrogen from the atmosphere and convert it to a fixed form that is available to plants. As legume roots break down over time, available nitrogen is released to the soil and can be taken up by other plants. In fact, a mixture of grass and legume requires no nitrogen fertilizer if the mixture is at least 33% legume on a dry matter basis. Environmental benefits specific to legumes within forage-based cropping systems include nitrogen-fixation along with providing high-quality nectar and protein-rich pollen for bees. Most legume species require insect-mediated cross-pollination and have coevolved with bees to provide the pollen and nectar as protein and carbohydrate, respectively, as rewards for pollinators (Fig. 13.1).

Fig. 13.1　Red clover with pollinator (photo credit: N. McLean)

In terms of animal feed, forage legumes have a high protein content, which is related to access to available nitrogen. Forage legumes also have a higher mineral content than grasses. Forage legumes often have taproots that can access soil moisture at greater depths than grasses, which provides greater yields when moisture becomes limiting. Forage feed quality declines as plants mature, but feed quality of legumes declines more slowly than with forage grasses.

While forage legumes have several advantages compared to forage grasses, there are different advantages related to grasses. Grasses have a fibrous root system that holds soil in place making it less vulnerable to erosion. While grasses have lower protein than legumes, they are less likely to cause bloat in ruminant livestock (*Bloat* is a form of indigestion marked by excessive accumulation of gas in the rumen). The combination of lower protein and lower minerals makes grasses easier to store as a silage, based on lower buffering capacity. Grasses dry more quickly than legumes after being mowed, which aids in harvest for stored feed. When forage is cut and left to dry in a field prior to baling for hay, legume leaves are prone to shattering while grass leaves are more likely to stay intact. Finally, perennial grasses often persist for a longer period than perennial legumes.

A mixture of one or more forage legumes with one or more forage grasses can make use of advantages of both plant families. As stated above, nitrogen fixation by the legume will also provide adequate nitrogen for the grass, thereby eliminating the need for applications of fertilizer nitrogen and the associated environmental and economic implications. It is important to match legumes and grasses within mixtures so that they are adapted to the local soil and climatic conditions and they reach optimal stage for harvest at the same time.

13.4   Harvesting and Storing Forage

Forage crops can be harvested in three ways: pasture, hay, and silage. Pasture is the least expensive way to harvest forage plants because animals harvest it themselves. Farmers must manage the crop and animals to optimize yields. Highest yields of forage plants and livestock result from rotational grazing where animals are fenced within restricted areas, then moved periodically to un-grazed areas. The forage plants are provided with a rest period to build up root reserves. Benefits of rotational grazing are greatest with high-yielding forage species. These species do not persist under continuous grazing. The longer that animals are kept on the same area, the greater the waste of plant material from trampling. Forage harvested by grazing animals has higher feed value than from other types of harvest because there are no storage losses and the animals may select higher quality material (Fig. 13.2).

Fig. 13.2   Cattle grazing on a pasture near Cape John, Nova Scotia, Canada (photo credit: N. McLean)

Pastures provide animals with the highest quality feed and at the lowest cost; however, pastures are not always available due to factors including heat and drought, excessive moisture, or cold winters. Pastures are also not feasible for dairy farms with limited pastureland near milking facilities. For a variety of reasons, most farms preserve some or all of their forage to feed at later dates. Forages are preserved like human food, thorough drying, as hay, or fermenting, as silage.

Hay is forage that has been cut, left to dry in the field to less than 15% moisture and is then stored, usually after being baled (Fig. 13.3) for ease of handling. Hay must be kept dry to maintain quality. The dry conditions prevent growth of microorganisms that could cause spoilage. If hay is baled too wet, then microbial activity reduces forage quality and releases heat that causes protein digestibility to be reduced, through the Maillard reaction and, in some cases, enough heat is generated to cause fires.

Fig. 13.3　Harvesting and baling hay in Nova Scotia, Canada (photo credit: N. McLean)

Silage is forage at 40%–80% moisture, depending on the type of silo, that is preserved by anaerobic fermentation. Lactic acid-producing bacteria multiply on the forage and release organic acids, which reduce the pH to a point where no further microbial growth occurs. Silage must be kept in air-tight silos to prevent degradation by aerobic microbes. Silos include plastic-wrapped baled forage, bunkers (with walls), or drive-over piles (without walls) covered with plastic, and towers constructed of concrete, wood, or steel. Type of silo is often related to the type and size of the farm. Smaller farms often produce wrapped bale silage while large farms often have bunker silos. Some farms have more than one type of silo.

13.5　Forages in Rotations

Forage crops are commonly recommended as components of crop rotations due to beneficial impacts on soil structure, soil nitrogen status, and soil organic matter. Many long-term studies have found that other crops, which followed forages in rotation, performed better, particularly when the forage contained legumes. Grain corn/maize (*Zea mays* L.) shows very significant yield increases when forage legumes are components of the rotation. Benefits in Iowa, USA, extended beyond what could be explained by increased soil nitrogen availability (Osterholz et al., 2018). A 19-year study in Ottawa, Canada, found that a simple annual rotation with 1 year of corn followed by 1 year of a forage legume (alfalfa or red clover) significantly reduced nitrogen fertilizer requirements and also reduced greenhouse gas emissions and carbon footprint compared to monoculture corn (Ma et al., 2012).

Potato (*Solanum tuberosum* L.) is commonly grown in rotation with forages and other crops in order to prevent soil degradation and subsequent erosion. A traditional rotation for potato in eastern Canada is potato in year 1, followed by *Hordeum vulgare* L. (barley) which is under-seeded with a perennial forage grass/legume mixture in year 2. Barley is harvested in year 2 and the forage crop is either incorporated into the soil or killed with herbicide in fall of year 3.

A 17-year study in Brazil (dos Santos et al., 2011), comparing six different maizebased crop rotations concluded that the highest level of carbon sequestration was accomplished when *Medicago sativa* (alfalfa) was intercropped with maize (2 years of alfalfa hay intercropped with maize every third year). The second highest C sequestration was from a rotation of winter *Lolium multiflorum* Lam (ryegrass) hay crop—summer maize—winter ryegrass hay crop—summer *Glycine max* (L.) Merr (soybean). Rotations without forage crops contributed less carbon to the soil.

A study in southwest China (Li et al., 2018) reported that soil organic carbon and total nitrogen were depleted following 15 years of a maize–soybean rotation compared to the native forest. Both soil organic carbon and total nitrogen increased significantly when a perennial forage grass, hybrid Napier grass [*Pennisetum americanum* (L.) Leeke × *Pennisetum purpureum* Schumach] was grown. Two other perennial crops in the study, mulberry and sugarcane, did not result in increased soil organic carbon or total nitrogen.

Summary

Forage crops are grown to be fed to livestock and they cover more land than any other type of agricultural crop. Forage crops cannot be eaten directly by humans; however, they can indirectly provide products for human use. Most forage crops belong to two plant families: Poaceae (grasses) and Fabaceae (legumes). Forage crops are often grown on land that is not suitable for production of fruits, vegetables, or cereal crops but can be converted by ruminants, camelids, or horses into products for human food or use. Additionally, forage crops provide a wide variety of ecosystem services including nitrogen fixation, carbon sequestration, food and habitat for wildlife, protection of waterways from run-off, and protection against soil erosion. Forage crops are harvested by grazing animals or by farmers in the form of hay or silage. Crop rotations often include forage crops for beneficial impacts on soil health.

Chapter 14
Cereal-Based Cropping Systems

C. D. Caldwell and S. Wang

> *"The economy is a wholly owned subsidiary of the natural ecosystem."*
> —Paul Hawker

Abstract The cereal grains are the most successful group of crops in the world. The big three are maize, wheat, and rice. Human society has evolved in a codependent way with cereal grains and the solutions to the hunger problems facing humanity in the next century revolve around how we will develop these crops further for the benefit of humankind.

Learning Objectives

After studying this topic, students should be able to:

1. Describe the reasons for the economic and human health importance of cereal crops.
2. Name the three principal cereal crops in the world and their major uses.
3. Describe the role of cereals in:

   (a) Livestock-based rotations
   (b) Cash crop systems

4. Describe possible strategies for improving these crops to meet future needs.

14.1 Introduction: Why Are Seeds of Cereals So Great?

World hunger is an overarching issue and shall be a major concern for the remainder of the century. To a significant degree, advances in crop production will determine whether the hunger problem can be solved or not. The amount by which the annual food production must be increased to meet the food demands and projected change in diet will greatly depend on improvements in crops and cropping methods under stresses of decreasing arable land, increasing soil degradation, and changing climate. Present estimates of the increases needed range from 2.5% to 5% annually. Research, development, and extension regarding the growing of cereal crops will be central to success in meeting those goals (FAO, 2018).

Humans and the major cereal crops have coevolved over the past 10,000 years. The domestication of cereal crops has provided the food and stability for the development of human society. Cereal crops provide the basic energy for almost everyone on Earth. In addition, these little kernels of energy also provide nutritive fiber and a wide range of vitamins and minerals. These domesticated plants have spread all over the world as humans have expanded our society. We need them and they would certainly not survive without us. The challenge will be to improve even further the ability of these few plants to support our society.

There are approximately 35,000 species of flowering plants, more than 3000 of which have been identified as economically important; however, we are dependent on 15 plants for almost all our food. Of those 15, 8 are cereal plants; Table 14.1 shows the world production of principal food crops; figures are in thousands of metric tons. Half the total global production of food is cereals; there are three major ones: maize, rice, and wheat, the latter two make up more than one third. The food supply for humans has a very narrow genetic base, a low diversity production base to support more than 7 billion people in the world. There is some considerable fragility around this. From an ecological perspective, stability of an ecosystem is directly proportional to diversity. We now have a problem in the world of a narrow crop genetic base in our food system.

Table 14.1 The ranks of world crop production (Index Mundi, 2020)

Rank number	Crop	World production (× million T)
1	Maize	1108
2	Wheat	762
3	Ric	513
4	White potato	388
5	Soybean	337
6	Cassava	277
7	Barley	156
8	Sweet potato	120
9	Oil palm	75
10	Oilseed rape	68
11	Sorghum	58
12	Sunflower	53
13	Groundnut	45
14	Pearl millet	30
15	Oats	22
16	Rye	11

Cereal grains are grasses (members of the monocot family, Poaceae), which are cultivated for the edible components of their seeds (botanically a type of fruit called a caryopsis), often broken down into its parts: the endosperm, germ, and bran.

As Table 14.1 demonstrates, of all the world crops, the grains of cereal plants are the most widely grown. They provide more food energy than any other type of crop and in general, they form the staple diet of most people on the planet. Raw kernels of cereal grains are not only a significant source of fats, oils, or starch but also provide important vitamins and minerals. Unfortunately, in refining some of our major cereal grains, both the bran and the germ are removed, leaving only the endosperm (Fig. 14.1). When that happens, all that remains is the endosperm, which is mostly carbohydrate and lacks other significant health-giving properties. If the diet is supplemented with other foodstuffs such as fruits and vegetables, the use of refined cereals is not a dietary danger; however, under poverty situations in which the rice, wheat, or maize constitutes almost all of the daily food intake, this loss of nutrients from the outer parts of the kernel is a significant health problem. In many developed countries, there is a significant amount of cereal grains fed to livestock.

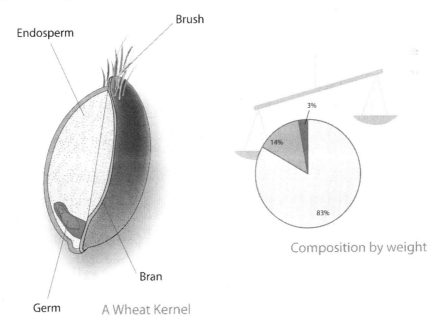

	Carb./g	Protein/g	Fat/g	Fiber/g	Iron (% daily req.)	Others
Bran	63	16	3	43	59	vitamin Bs
Endosperm	79	7	0	4	7	
Germ	52	23	10	14	35	vitamin Bs omega-3/6 lipids

Nutritional Value (per 100g)

Fig. 14.1   Wheat-kernel composition (https://commons.wikimedia.org/w/index.php?curid=12889006)

Cereals, in general, are not high in protein (usually in the range of 10%–15%) and the amino acid composition of the protein does not completely match the needs of people. Soybeans and other grain legumes have protein content higher than that of rice or wheat and the amino acid composition complements that of the cereals. The protein content of raw soybean is around 35%. Soybean meal, which is used mostly for animal feed, is around 44% protein after processing. A lot of protein in the world comes from legumes but, unfortunately, not all people can afford basic food beyond the cereals and these low-protein crops like rice, wheat, and barley provide about 50% of the protein consumed by people worldwide, despite being a lower form of protein.

For humans to use the amino acids in food, we need a full complement of amino acids at the same time from different sources. A prime example of such protein complementarity is the practice of eating rice and beans together. The amino acid composition of cereals and legumes complement each other and provide more human nutrition than either would when eaten separately. The human body takes a protein from a plant and breaks it down into amino acids and then rebuilds it into our own proteins. Our protein and animals' protein have different ratios of amino acids than that of plants; e.g., maize is short in lysine. Therefore, we can consume lots of protein from maize but its usefulness will be limited by how much lysine there is in the maize in order to make our own protein. There will be many "extra" amino acids that we cannot utilize because the lysine is limiting. One can get protein deficiency even though they are taking in a good amount of protein. In a fair and sustainable system, humans should not be relying on cereals for their protein.

14.2　Domestication and Adaptation of Cereals

The temperate cereals evolved around the Tigris and Euphrates rivers in the area of what is now Iran and Iraq. This area is therefore the site of origin and where one might find the greatest species (genetic) diversity for those crops. When the cereals were first selected by hunter-gatherer people, they would have simply been mixed species. Under very good conditions, wheat tends to outcompete the barley and the oats but as people moved further north and the prevailing climate changed, the barley and the oats became more dominant. As they moved with their mixture of harvested seeds, there was natural selection taking place across the world. The species sorted themselves out across Europe in the places they grew best. The first selections were made by accident as people moved. Some types were better suited to certain areas. Since that time, of course, people have done a lot of targeted selection; some seed stored well and those that did not store well were not planted. The ones that were bigger were planted more and yielded more, so essentially a lot of selection for these types of crops occurred thousands of years ago. We have done a huge amount of cereal improvement over the last 100 years, but we started with well-established species through selection by our ancestors. Present-day plant breeders now select and breed for several desirable characteristics, e.g., larger seeds for harvest. This is true for several crops but particularly true for the small grains. We want bigger grains and we want to harvest easily as well; in other words, we want them to stay on the cob or head until we come along with a combine or other harvest method and thresh them. We do not want the seeds to fall off; we do not want seeds to shatter as they get ripe. Shattering means that as soon as seeds are ripe, they fall to the ground; this is a survival characteristic of wild plants that have been bred out in our domestic types. We breed for ease of harvest, no shattering, and less toxicity. Most wild plants have some sort of chemical protection. This chemical protection is quite common among wild plants and is stimulated by stress, especially predation by insects. With any stress, the plants tend to produce more of that protective compound. For example, wild potatoes have a lot of toxins. Even with present domestic types, we do not eat the leaves of potatoes because they are still toxic even though the tubers have had the toxin bred out of them. That same toxin that is in the leaves is in the roots and in the tubers of the wild precursors of our domestic potato and we have bred it out. In order to domesticate plants, we have tended to create bigger organs, easier harvest, and less toxicity. In addition, lately we have done more breeding to develop targeted substances within the kernel. For instance, we now have designer oils being bred into canola (oilseed rape). Now we are breeding not just for cooking oil but rather breeding for something of a higher value that can be extracted and perhaps put into a little capsule to be sold at a premium price.

14.3   The Major Cereal Crops in the World

The total world production of cereals is just less than 3 billion metric tons per year (FAO, 2020). More than half of the total cereal production globally comes from only five countries; in order of importance they are China, the USA, India, the Russian Federation, and Brazil.

14.3.1   Wheat

Wheat (Fig. 14.2) was probably the first of the cereals to be domesticated and cultivated by humans, perhaps as long ago as 18,000–12,000 B.C. Wheat is the world's most widely grown crop (area wise) and together with rice, these two crops share almost equally in fulfilling the energy needs of people. Wheat is harvested somewhere in the world every month of the year; it is cultivated as far south as a tip of South America and as far north as Alaska and Finland. However, it is best suited to the temperate latitudes. Wheat is not well adapted to tropical or semitropical conditions. Worldwide, wheat is a cereal of international trade; a high percentage of all wheat that is produced crosses an international border before being consumed. Wheat is grown on all continents and is the most important cereal crop in the Northern Hemisphere, Australia, and New Zealand.

Fig. 14.2　Wheat influorescence

(photo cridit: C.D. Caldwell)

14.3.2　Rice

A literal translation of the Chinese daily greeting is "have you eaten rice?" (吃饭了吗？) The inference here is if you have not had rice, you have not eaten. Traditionally in China and many of the countries in Southeast Asia, one would have rice three times a day. Rice is an integral part of the history, culture, and economy. Rice is the most important food crop in the world; it is a primary source of food for more than half the world's population. Rice production is largely concentrated in Asia; more than 90% of the world's rice is grown and consumed in Asia where 60% of the Earth's people live. It is planted on about 154 million hectares annually or about 11% of the world's cultivated land. In China, the percentage is even higher; 25% of China's cultivated land is in rice. Unlike wheat, rice is a cereal of sustenance, not principally a cereal of commerce. Most of the rice consumption occurs within the same region in which it is produced (Fig. 14.3).

14.3.3　Maize (Corn)

Maize (Fig. 14.4) is the only major cereal which originated in the Americas. Many types of maize have evolved and been selected over thousands of years for adaptation across the globe. More than any other cereal, massive breeding efforts in modern agriculture have gone into maize to narrowly adapt it to specific regions. Today maize is a staple food for a large portion of the population in the world. It is especially important to the diets of many countries in Africa, Central and South America. Maize, however, has many applications beyond the food market. In North America, more than 30% of maize is used for industrial production for many products as diverse as alcohol and plastics (CIMMYT, 2019).

Fig. 14.3　A rice field in Ninghua County, PRC in 2017 (photo credit: S. Wang)

Fig. 14.4　Maize in the field (photo credit : C.D. Caldwell)

The USA remains the largest producer of maize in the world. However, China is now the # 2 producer of maize. Over the past several years, maize has started to replace rice as a top crop in the country. This is not due to change in the dietary patterns of the Chinese regarding rice; rather it is the increasing demand for feed for livestock in the country that is driving the increase in maize production. With increasing urbanization, greater wealth, and tastes for a Western lifestyle, meat consumption has increased dramatically. This has meant an increased demand for livestock feed, especially maize.

14.4   Cropping Strategies with Cereals

Cereals are not always grown as a cash crop. Sometimes, the cost of production for small grains can be almost equivalent to the returns; in some areas of the world, especially under small farm conditions, it is difficult for a grower to make substantial returns on small grains. In such cases, they may be more beneficial as a rotational crop. Cereal crops have fibrous roots that hold the soil well and have good characteristics in terms of breaking disease, insect and weed cycles that are advantageous in many high-value crop rotations. They are rotated around some other cash crops that produce more income. Most cereals have both spring types and winter types. Winter types are winter annuals; they are planted in the fall, go dormant over the winter, and are harvested the next summer. It is an annual because it completes its life cycle within 1 year but it is a winter annual because it is planted in the winter and harvested the next summer. In temperate climates, having a winter cereal giving soil coverage rather than a bare soil is of great advantage to prevent soil erosion. Spring cereals are planted in the spring and harvested in late summer and provide no winter cover.

As was discussed in Chap. 6, cereals are key components in rotations for providing ecosystem services. In this case, the ecosystem services come from a rotational role of the cereal. What does a farmer want from a rotational crop? Instead of making money directly, the purpose of the use of cereals in a rotation is to maintain soil integrity, good soil health, good organic matter, efficient use of inputs, and reduce erosion. Keeping the ground covered, it gives a direct economic return to the growing year in terms of sustainable agriculture. Cereals are rotated with crops like potatoes, carrots, and other vegetables that can make money and break the disease and weed cycle. For example, the diseases and insects that affect potatoes do not affect cereals, thus decreasing the use of fungicides and herbicides. Cereals can also act as a nurse crop for forage crop establishment. For instance, a standard rotation in potato growing areas is grain under-seeded to some forage, such as red clover/Timothy mix. During the year of the cereal production, the forage establishes, grows a bit and, when the cereals are harvested, the forage will flourish. The forage can be harvested for animal feed the next year or plowed down to improve soil health; the result will be improved soil structure and a disease cycle of the potato will be broken. In some places, such a three-year rotation has even been legislated by law in order to protect the soil. One of the side effects of such a three-year (or more) rotation is to improve the quality of vegetables such as potato. The improved soil conditions make for better growing conditions for the roots or tubers or whatever vegetable is grown.

The same sort of thing happens with carrots, strawberries, or other horticultural crops; cereals break disease cycles. They also add value in animal products so the farmer might not make money on the feed but if he/she can grow their own feed, that decreases the cost of production of the animal, thus making the money through the sale of the animal.

In short, there are people who use cereal crops for direct economic return; however, cereal crops are used even more extensively in rotations for their benefits. If one is a cereal crop producer, there are a few ways of making money: direct sales, indirect sales of animal products, and enhancement of value to other crops in the rotation. Perhaps the highest return a farmer can get for any of these cereals is to grow seed and become a certified seed grower. If a farmer is producing cereal crops, there is an opportunity to grow part of their crop area for sale as seed. The equipment is available and the farmer just needs to develop enough expertise to grow seed effectively. A quality pure, weed-free, and disease-free seed gets a very high return. The farmer must be very good at growing the grain and producing a clean product with no contamination of weeds, no disease. For direct sales, seed is the highest value, milling is the next for wheat, malt for Barley, human use milling for oats. The lowest direct return for cereal crops is for animal feed. All these products are graded so the higher the grade, the better the return. Interestingly enough, there are years when the straw from the cereal crop attracts more money than the grain itself because there is a demand for straw for some years for spreading on land to prevent erosion, to put on horticultural products such as strawberries or for use by mushroom growers, so there are various other opportunities.

From the farmer's point of view, the idea is to aim for the highest return market, a known market and to produce a quality grade, while at the same time maintaining or improving the state of the soil and the environmental sustainability of the farm. Most horticultural growers do things in a timely fashion in order to realize the highest return. In the same way, a grain grower must do things in a timely fashion or the result is poor quality, low yields, and poor economic and environmental impact.

14.5   Strategies for Improving Cereal Crops to Meet Future Needs

As mentioned above, there is a critical need to increase production of our key cereal crops in a sustainable way in order to meet the nutritional demands of the future. From a plant breeding standpoint, this will mean developing genetically diverse, high yielding new cultivars of not just the big three cereals mentioned above, but also some of the other very important cereal grains. These include sorghum and millets which are key food grains, especially in some of the less developed countries of the world.

The breeding goals need to combine tolerance to multiple stresses, especially heat and drought stress in response to climate change; efficient nutrient use; improved nutritional quality; and adapted seed for targeted production areas. In addition, resistance to major diseases, insect pests, and parasitic weeds cannot be downplayed. Historically, breeding efforts have concentrated on serving larger farmers who are able to pay for the improved technology of new seed. We also need to develop through public money for public good, adapted seeds for production areas where small farmers are the dominant producers. From an environmental sustainability consideration, some of this breeding effort will be targeted toward increased root mass for soil retention and water scavenging, bringing both economic returns as well as environmental efficiency.

Because of the need to increase the genetic potential of the cereal grains quickly and safely, it will be necessary to use all the tools in the breeder's toolbox, working in concert with regulators and governments. There needs to be an integrated application of modern tools and technologies, but not in isolation from cropping systems and food and value chains. Development and testing of new lines must take into consideration and cooperation of farmers themselves.

On a local level, capacity must be developed in appropriate cereal breeding and seed systems, especially with women farmers and the youth so that there is a continuous system of training and knowledge exchange. This means that there must be a crucial link between scientists, breeders, and community in order to accelerate the rate of change and adaptation for new genetics.

This cooperative work among scientists, extension workers, and the community can result in science-based recommendations and improve productivity on a regional basis. In short, strategies for the future must be ecosystem-based, community-based, future focused, and adaptive to change.

Chapter 15
Vegetable-Based Production Systems

C. D. Caldwell and S. Wang

"They say that vegetable food is not sufficiently nutritious. But chemistry proves the contrary. So does physiology. So does experience... the largest and strongest animals in the world are those which eat no flesh-food of any kind - the elephant and the rhinoceros."
—Russell Trail

Abstract Vegetables are a key component of both the economy and the human health. Vegetables are increasing in production throughout the globe with a wide diversity of types reflecting both culture and climate. Asia dominates the production and export market of vegetables with a large diversity of types. High economic return from vegetable growing has stimulated some unsustainable practices. It is important for the future that ecological principles are used to reinforce the need for cropping systems using vegetables to make money but also sustain the environment for the long term.

Learning Objectives

After studying this topic, students should be able to:

1. Discuss the importance of vegetables for the economy and for human health, in both a global and a local context.
2. Name the four most commonly deficient nutrients in human diets and describe the impact of those deficiencies.
3. List and briefly explain four ways of classifying vegetables, giving examples for each method of classification.
4. Describe the predominant vegetable production systems in the tropics and temperate climates.
5. Explain the profit vs principle concepts of vegetable growing.

15.1 The Roles of Vegetable in Human Development

Why do people grow vegetables? Vegetables are an important component in agricultural systems;

they are important for rural diversity, rural sustainability, and human nutrition.

Vegetables are the major source of energy and several essential human nutrients. When one compares rice to potatoes, about 40 times more calories can be harvested per hectare with potatoes. There is a great diversity of nutrients that exist in vegetables not found in cereals or fruits. They provide iron, Vitamin A, dietary fibers, proteins, essential amino acids, and other medicinal properties. Vegetables are consumed not only for their basic nutrition and energy but also for their overall health-giving properties.

Vegetables have high productivity; compared to 2–7 tons/ha for rice, carrots can provide up to 70–90 tons/ha per year. They can also produce about two to three crops per year depending on where they are grown. Vegetables are of higher value in terms of productivity and profit, anywhere from 40% to 300% compared to other crops. They also provide higher employment opportunities; it is estimated that vegetables provide about a four- to five fold increase in employment opportunities by switching over from a rice-based system to a vegetable-based system. In order to grow and commercialize vegetables, people need specialized skills such as training in pruning, propagation, nursery, and seed production; all these industries can be amalgamated together for a successful vegetable industry.

Vegetable production systems provide high diversity; unlike a cereal monoculture, one can have multiple crops, intercrops, providing biodiversity both in the terrestrial atmosphere and in the rhizosphere. Vegetable systems are also industry oriented; many value-added products and commodities come from vegetables. The seed, nursery, packaging, and processing industries all interlink and add value to a successful industry. At present, there are also efforts to deliver nutraceuticals and biopharmaceuticals through vegetable systems.

Also important is the concept that "vegetables can heal". Vegetables contribute to both prevention and curing of human diseases. There is a list of benefits that are available by consuming vegetables. Vegetables can provide antioxidants, antibacterial and antifungal agents; they can help immunodeficiency disorders. A diet with sufficient vegetable intake has been shown to have positive anticancer properties. Diets high in vegetables have been shown to have positive effects against cardiovascular disease function and lower blood cholesterol levels.

One of the major components of onions is allicin; there is evidence of antiviral, antifungal, and antibacterial properties related to this compound. A tropical vegetable with apparent medicinal properties is bitter melon (Joseph and Jini, 2013); there is evidence that diabetes (type A) can be positively affected this vegetable. Research continues on how various vegetables have both nutritional and medicinal properties.

15.2　Nutrient Deficiencies

The World Health Organization reports on the four principal constituents that are deficient in the diet of most of the population in the world; they are protein, iron, iodine, and Vitamin A. Many people are familiar with protein deficiencies; however, most are less aware of the effects of the deficiencies in the other three principal constituents.

Iron deficiency is the most common nutritional disorder in the world. It affects as many as two billion people; over 30% of the world's population is anemic due to iron deficiency, the health effects of which are exacerbated by infectious diseases. The health effects globally due to iron deficiency include poor pregnancy outcomes, impaired physical and cognitive development, increased risk of other diseases in children, and reduced work productivity in adults. Anemia is a contributing factor to 20% of all maternal deaths. Southeast Asia is one of the chronic areas suffering from anemia due to lack of iron in the diet; about 78% of the population is anemic.

Iodine deficiency is the world's most prevalent, yet easily preventable, cause of brain damage. The effects of iodine deficiency can be severe. Such deficiency during pregnancy can result in stillbirth, abortion, or mental retardation. Lower levels of deficiency often result in mental impairment, which reduces intellectual capacity at home, in school, and at work. This insidious deficiency therefore can decrease productivity and increased poverty.

Vitamin A deficiency (*VAD*) can cause blindness in children and increases the risk of disease and death from severe infections. In pregnant women, VAD can produce night blindness and increase the risk of maternal mortality. Recent statistics show that 250,000 to 500,000 vitamin A-deficient children become blind every year, and half of them die within 12 months of losing their sight. This is a serious public health problem in more than half of all countries and is especially prominent in Africa and Southeast Asia.

All four of these key constituents can be supplied by vegetables. For example:

PROTEIN:	Lima bean, green pea, kale, broccoli, sweet corn
VITAMIN A:	Carrot, sweet potato, spinach, beet greens, chard
IRON:	Spinach (cooked), peas, turnip greens, asparagus
IODINE:	Green beans, lima beans, chard, turnip greens, (iodized salt)

15.3 Socioeconomic Impact

In terms of the health effects of vegetables and their consumption needs, the World Health Organization recommends about 73 kg per person per year for the intake of vegetables (WHO, 1990). Several countries are near or just under the line (Fig. 15.1). The world population in February 2020 is recorded at 7.8 billion people, increasing by a rate of 80 million people per year. The major population increases over the next 20 years will happen in parts of Asia and Africa. What will happen in that situation when migration is taking place from the country and rural areas to urban cities? It is estimated that 26 million live in the urban cities of Shanghai and 20 million people live in Mumbai which means there needs to be a regular supply of vegetables to be moved into the urban areas. This need will only increase in the future.

Fig. 15.1　Per capita vegetable availability of Asian countries in 2000

Because of their importance to human health and their economic influence, vegetables have a social and economic impact. There is a continuous connection between consuming vegetables, producing vegetables, and building a nation. It is not just the vegetables we need to produce but it has other sociopolitical implications in building countries, building continents, building regions, building the whole global situation. For instance, improved nutrition can lead to improved health, it can trigger learning capacity, and improve working capacity, the whole efficiency of a country can be enhanced by improved nutrition, thus improving competition among different countries, and enhancing socioeconomic development. The improved socioeconomic development could result in improving the infrastructure and institutional development. It will contribute to increased income and ultimately that will serve as grounds for improved nutrition. The cycle becomes positive not negative.

Vegetable Production: Global and Regional

In 2019, sales of vegetables worldwide generated more than $1.3 billion, an increase of more than 3% from 2018. There is an upward trend in market value increase throughout the period from 2005 to the present, with some marked differences from year to year. In general, global vegetable consumption is increasing and predictions are that it will continue to do so. Production is increasing and trade, both nationally and internationally, in vegetables is increasing.

As can be seen from Table 15.1, most of the vegetables in the world are produced in Asia; of the countries in Asia, by far the most vegetables are produced in China (554 million metric tons per year) and India (127 million metric tons per year). By comparison, the USA produces only 32 million metric tons per year. China is also the world's largest exporter of vegetables.

Table 15.2 demonstrates the most important vegetable types grown globally. Note that the number 2 group, brassicas, includes a great number of different related vegetables including such diverse ones as cauliflower, broccoli, and cabbage.

Table 15.1   Global production of vegetables in 2017 in millions of metric tons (FAO, 2018)

Region	Total production in 2017	Region	Total production in 2017
Asia	834	Africa	79
Europe	96	Oceania	33
America's	81		

Table 15.2   Production in 2017 of the top 10 vegetable crops annually in million metric tons (FAO, 2018)

Vegetable type	Total production in 2017	Vegetable type	Total production in 2017
Tomatoes	182	Carrots and turnips	42
Brassicas	97	Chilies and peppers	36
Onions	93	Garlic	28
Cucumbers	83	Spinach	27
Eggplant	52	Pumpkin and squash	27

15.4   Classification Systems of Vegetables

The term vegetable is not a particularly scientific one. In essence, it may be considered a culinary term wherein it is defined as any plant part that we ate as part of the main meal. Obviously, this is a very vague definition. However, how we define vegetables is determined by the purpose of that definition. Breeders of new types of vegetables will have certain ways of defining the types of vegetables; plant physiologists and teachers of plant anatomy and growth may define vegetables in terms of their plant growth or even their environmental requirements; most of us, however, actually define vegetables on the edible parts that are consumed.

Therefore, we find that vegetables are defined in four most common ways: Vegetables in many ways, these are the four most common:

- Botanically
- By the edible parts that are consumed
- By their temperature requirements
- By their life cycles (culturally and botanically)

The only universally accepted classification system for crop plants is botanical classification. This involves classifying groups of plants into kingdom, division, subdivision, phylum, subphylum, class, subclass, order, family, genera, species, subspecies, and variety. In discussing vegetable classification at the botanical level, we are usually referring to the level of family. Each crop plant is also designated with a scientific name (in Latin), which includes the genus and species that are accepted worldwide. While academics find this useful for communication, it is of little use to the farmer. As an example of this, botanists will know that both potatoes and tomatoes are closely related botanically; however, their growing condition requirements are quite different. Knowing that they are related botanically does not help a producer know whether they should grow them under the same growing conditions. However, what it does tell a farmer is that, because they are related botanically, they may also host the same insects and diseases and should be separated in any rotation. By contrast, on a botanical level carrots and radish are very different, yet have the same types of growing conditions.

There are about 200 species of vegetables consumed worldwide. All vegetable crops belong to the Angiosperm division. Most vegetables we consume come from two different subclasses of angiosperms, monocots, and dicots. Vegetables are not from any one species or variety; they are a diverse type of food. Botanical classification gives an understanding of how they are assembled in terms of their anatomical characteristics; The classification is based on plant structure and similarities and dissimilarities in cellular organization, where they originally evolved, what other crops they can cross with, type of flower produced. Structural relationships are a result of evolution, e.g., the Cucurbitaceae family have similar morphology in terms of the leaves and some of the reproductive organs and the Polygonaceae family have similarities in the morphological and reproductive characteristics. Botanical classifications are especially important to vegetable breeders, since they give the framework on which to look for desirable traits that can be used in crossing. Botanical classifications are of use but are not the only way that we classify vegetables. It is the user of the information that often determines the best way to classify.

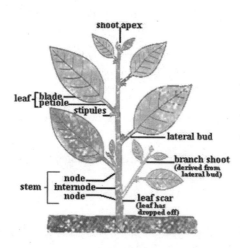

Fig. 15.2　The edible parts in vegetable plant shoot

Vegetables can also be classified based on *edible parts*. There is a whole spectrum of plant organs we consume; i.e., roots, tubers, stems, leaves, petioles, inflorescence, and immature and mature fruits (Fig. 15.2). Radish is a root, we eat the tuber from a potato plant, onion is a bulb, broccoli is an immature inflorescence, and tomato and watermelon are botanically classified as fruits. There are also leaf modifications, for example, brussels sprouts.

Vegetable varieties can be selected for production in certain climates (e.g., tropics) and growing season (e.g., cool season) according to *their temperature requirements*. Cool-season vegetables include asparagus, broccoli, brussels sprouts, cabbage, and kohlrabi; warm-season crops include cucumber, eggplant, and tomato that can be planted in tropical area even in summer, whereas, cool-season crops can only be grown in cool temperatures. This gives an option to growers, especially in warmer climates, as to which crops can be grown at specific seasons.

Vegetables can also be classified based on the *life cycle*, i.e., annual, biennial, and perennial.

Annual crops like cucumbers, cowpeas, and lettuce will complete their life cycle and therefore their edible part, in one growing season. Biennial crops have two cycles; they have a vegetative cycle and then enter the reproductive cycle the second season. Some biennial vegetables are grown as annual crops as the vegetative portion is the end product not the seeds, e.g., carrots. Seed producers of these crops would require the 2 years for the seed production. Perennial vegetable crops are few; asparagus is an example of a herbaceous perennial vegetable that does not have to be replanted each year and the new spring shoots can be harvested year after year.

15.5   Vegetable-Based Production Systems

There are two major categories of production systems, tropical and temperate (Fig. 15.3). There are subcategories or classifications by cereal-based system or year-round production systems, which means the continuous vegetables producing within a year-round on a specific piece of land versus rotation with cereals and other crops within a year-round sequence. There is also a mixed system; e.g., ten months' cassava in crop duration, with other vegetables planted along with cassava. A yearround production system has two major classes of categories, cultivated in a field or in a greenhouse or protected structure system.

The *perennial* vegetable system is common for both tropical and temperate areas. The vegetables are planted once; an example for the tropical systems is Moringa; it has long pods of edible fruits as well as leaves that are fed to livestock. They contain a high amount of iron; it is in many backyards of every country in any tropical region. Coccinia is another example of a tropical perennial. It can be grown as a climber and the leaves and fruits can be used. In the temperate region, examples include asparagus and rhubarb. One of the great advantages of perennial vegetable crops is they can be planted, managed, and then harvested without replanting. In a cereal-based production system, rice or wheat (also barley or oats) are used as rotation crops with the vegetables in a tropical or temperate climate. The yearround production system produces vegetable after vegetable mostly in a tropical production system.

In a *temperate cereal-based* production system, the wheat or barley is followed by potatoes or carrots or any of the crucifer crops, and corn, cauliflower, cucumbers. In a forage-based system, the forage may be followed by onions, watermelons, or tomatoes, depending on the market trends.

There are many different cropping systems that are available in a temperate vegetable production; e.g., combinations of crops like potato, cereals, spinach, and cereals or forages in a specific system followed by a vegetable.

Highland vegetable rotations: In some of the tropical countries, the elevation is utilized as an advantage. They also grow vegetables in high elevations where the temperature is generally lower. In principle, it is a potato-based system with potatoes, tomatoes, cauliflower, or any other crucifers followed by peas, carrots, and onions.

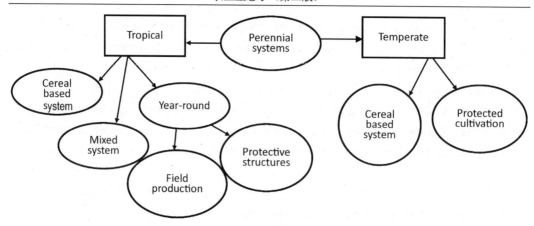

Fig. 15.3   Global vegetable production systems

Tropical production systems are often cereal based; some are cash crops. There are more than 2000 different production systems related to vegetables in the tropics. It is the experience, the market demand and the industry that determines what kind of cropping system happens within a certain rotation. Examples of a rice-based system would be rice–cucumber/watermelon/beans–rice; rice cucurbitaceous vegetable–rice; rice–cotton–tomato, a high cash crop system; rice–tomato–rice; rice–chilies–rice. In dryland areas, millets are drought-hardy cereals which will replace rice or maize in the rotation and you can have millets–chilies–millets. There are two major categories that are purely vegetable-based systems or cash crop-based systems. As the name implies, a purely vegetable-based system includes just vegetables while a cash crop system may include other crops like chilies, cassava, or bananas. It depends on whether the farm is supplying vegetables to a contract vegetable supplier, a wholesaler or retailer, or if it is dealing with multiple industries.

There is a great opportunity for diversity in the vegetable cropping systems. The year-round production system is very intensive and can happen in an open field condition or in an enclosed system. Both are commonly seen in the tropics and in temperate climate conditions. The choice of a crop is determined by the market and the choice of the grower. Vegetables can also be grown in *an aquaponic system*, in which leafy vegetables are grown in this aqueous system with about 12 crops per year. The vegetables are produced in a fertilized floatation tub; any type of vegetable can be produced in this way, tomatoes, melons, okras, anything of value. If there is a problem with a piece of land, very intensive cultivation can be done using this system.

Vegetables may also be grown more intensively in *protective structures* such as greenhouses, high or low tunnels, or net houses. The most profitable crops identified by growers for such high-cost systems are tomatoes, cucumbers, peppers, cherry tomatoes, bitter melons, and lettuce. Net houses are used to prevent contamination by major pests and to utilize the natural rain for irrigation.

An *off-season production* system may be used to capitalize on the off-season higher prices. This happens mostly in the tropics. When the temperature goes up (April to September) followed by heavy rains, regular production comes down dramatically and prices go up. In order for farmers to utilize a specific time frame and get more profit, they go for off-season vegetable production system overcoming the constraints. Profits can be possible in summer if winter crops, such as cabbage and cauliflower, are produced in summer. If a farmer produces tomatoes in summer in the tropics, he gets a higher price. They have an option to go into production of vegetables during the off-season when there is a low availability or prices are high. The major problem is high rainfall; flooding is commonly seen from April to September, typhoon, heavy rainfall, heavy winds as well as root diseases. To overcome the problems, grafting with resistant rootstocks that can tolerate flooding and high temperature and constructing a shelter to protect against the wind and rain produce good growing conditions. Identification of such resistant rootstocks can also help plants grow better under flooding conditions, withstand high temperatures and bacterial wilt, and avoid root nematode damage.

The use of management techniques that diversify cropping systems frequently include inter-cropping, multiple cropping, and cover crops. *Intercropping* can produce short-term crops under a canopy, while multiple cropping grows several crops together for harvest over a period of time. In relay cropping, one crop succeeds another in a short period of time. Trap crops are utilized to attract certain insects away from the main crop in the trap rows so that the other crops can be protected. An example of that would be a peanut crop that can be used as a cash crop, and castor bean crop can be used as a trap crop. Cover crops are used extensively in both tropical and temperate conditions to avoid erosion problems and avoid the damaging effect of direct pounding rainfall on the soil surface.

15.6   Profit Versus Principle

Because of the potential for high returns from growing vegetables, it is quite common for the vegetable industry is be very intensive. Unfortunately, many farmers tend to grow one vegetable continuously season after season to pursue high economic value so the land will be highly productive. This management technique will soon exhaust and deplete the land and cause environmental problems. Obviously, intensive vegetable production systems producing continuous crops of one species or closely related species is not good for the soil or the environment; also using excessive amounts of any input such as fertilizers or pesticides needs to be avoided. However, excessive fertilizer use and land tillage are commonly seen in many intensive vegetable operations, especially in tropical situations. Soil testing is not commonly practiced, so farmers may put on more fertilizers than is required simply as insurance against loss yields. This practice results in no opportunity for the soil ecosystem to rebound from this intensive use. The long-term consequences are low fertility, low organic matter, increased compaction, and low microbial activity. Some soils have almost zero microbial activity because of intensive cultivation. Throughout the rhizosphere, microbial biodiversity is decreased in such an intensive system. Short-term profit sometimes overrides the need to have solid production principles, which would provide for long-term sustainability and profit.

High pest disease and high amount of pesticide use are also commonly seen, which is a consequence of such economy-based intensive systems. Land erosion may be a particular problem, especially in highland areas. Because many vegetables require sustained water use throughout their growing, low water tables due to overuse of irrigation water in many parts of the world are now becoming a major limiting feature for vegetable production. Increasing land degradation is commonly seen in such profit dominated systems and soil is becoming degraded. Once the soil has become degraded, it may take years to rehabilitate it. Any production system must take this into account; abuse of the soil resource cannot be sustained. Caring for the soil, which is the basis upon which healthy vegetables are produced, is the fundamental process for successful production.

Therefore, it is extremely important to consider carefully basic ecological principles that will protect and build the soil resource, maintain biodiversity, and not waste or pollute the water resources. It is essential to maintain appropriate methods such as crop rotation, green manure crops, and inclusion of leguminous crops in the rotation. It may be necessary in the short term to compromise on some profits; however, this will ensure a productive vegetable system in the long term. Farmers can maintain a sustainable rotation of vegetables, sustainable both economically and environmentally, by avoiding the sacrifice of production principles for short-term profit. Ecological principles must win over short-term profit.

Chapter 16
Perennial Fruit and Nut Production Systems

K. Pruski

"There is no magic money tree."
—Theresa May

Abstract The incorporation of perennial crops into agricultural systems has huge potential for positive impact both environmentally and economically. Long-lived, woody perennials that produce edible fruits and nuts provide diversification of nutrition, high-value raw and processed products and stable, undisturbed soils. However, there is a need for more research to quantify outputs, inputs, costs, and benefits, in order to find the correct balance between environment and economics both in the short term and in the long term. This will involve appraising both simple and complex approaches to incorporation of perennials to complement both annual and livestock-based systems. We need to consider carefully agroecological principles and practices as they apply in both small landholder and larger farm systems.

Learning Objectives

After studying this topic, students should be able to:

1. Describe the economic and ecological importance of perennialization.
2. List the major fruit crops in perennial agroecosystems.
3. List and briefly describe major environmental considerations involved in the production of perennial fruits and nuts.
4. Briefly explain major marketing factors for perennial fruits and nuts.
5. Describe what is meant by "value-added" products with examples of such products made from fruits.
6. Using the case study of almonds in California describes possible positive and negative aspects of perennialization.
7. Describe the basic tenets of agroforestry.

16.1 Perennial Woody Plants in Agroecosystems

Since the end of World War II, in order to meet increasing global food demands, the major strategy in agriculture has been to focus on maximizing crop yield, unfortunately in many cases to the detriment of ecosystem services (Foley et al., 2005). These approaches usually include simplifying systems; i.e., utilizing agricultural systems that are dominated by a single crop (monoculture) or a very simplified rotation using few crop species, almost exclusively dominated by annual crops over perennial ones. This strategy has resulted in great success in increasing total food production (MEA, 2005). However, this high level of productivity is dependent on large external inputs, including synthetic fertilizers and pesticides, the cost of which is subsidized by the use of low-cost fossil fuels. Maintaining high production often has meant wasteful levels of irrigation (Tilman et al., 2002). By contrast, more complex ecosystems that include diverse plant communities are key to the provision of ecosystem services as described in Chap. 6. One method for making agricultural systems more diverse and more beneficial in terms of agroecosystem services is to incorporate perennial crop plants either within the rotation or as part of an intercrop.

The loss or deterioration of ecosystem services due to oversimplification of agricultural production demands change. Agriculture must take on more of a multifunctional approach to ensure that more than just basic needs are met (Boody et al., 2005; Schulte et al., 2006; Swinton et al., 2007). One promising approach for diversifying agricultural landscapes is through "perennialization". This involves, incorporation of various perennial plants (including fruit and nut crops) as essential components of agroecosystems. This will only be successful if the chosen perennial crops are well targeted to both the growing conditions and the market available. This means taking careful consideration of all aspects both biological and economic. When making a comparison with simplified cropping systems, the use of various perennial plants has been shown to enhance water supply regulation (Gerla, 2007), water quality (Duchemin and Hogue, 2009), carbon sequestration and storage (Zan et al., 2001), soil quality (Moonen and Barberi, 2008), beneficial organisms for pest control and pollination (Fiedler and Landis, 2007), and biological functioning (Fonte and Six, 2010). Many countries have seen additional benefits beyond the biological; rural beautification and enhancement of tourism have been by-products of the incorporation of such diverse landscapes that include perennial crops (Milestad et al., 2011). In some constituencies, sponsored by climate change advocates, the introduction of perennial crops has been promoted for carbon sequestration (Asbjornsen et al., 2014). For many recent decades, agronomic research has been focused on the development of technologies to maintain high crop yields under various climates (Gregory and Ingram, 2000; Motha and Baier, 2005). However, these approaches have not always taken into consideration the importance of sustainability of the diverse ecosystems in agricultural landscapes (Smith and Olesen, 2010). Increasing use of perennial crops, including woody perennials, increases biodiversity, and ecosystem stability and resilience (Lin, 2011; Tilman et al., 2006; Jackson et al., 2007).

Although the benefits of using perennial crops are very clear, it is not known specifically how much of the overall agricultural landscape needs to be of a perennial nature in order to benefit fully from the ecosystem services without jeopardizing the ability of the agricultural system to provide food and fiber. We require more research into aspects of the most cost effective and ecologically sound incorporation for perennial crops both spatially and temporally within a production system. However, whatever the conclusions of such studies, fruit crops, and nuts are already known to be perfect additions to perennialization of ecosystems.

16.2   Economics and Health with Fruit and Nut Crops

Fruits are a source of vitamin C, beta-carotene, potassium, polyphenols, and many, many more compounds. Nuts are an energy-rich source of protein and unsaturated fats. Both are excellent sources of dietary fiber and both can be used as ornamental trees in landscaping.

The world's major fruit crops are the tree fruits—apples, bananas, mangos, oranges, peaches/nectarines, lemons/limes, and pears and the small fruits—grapes, strawberries, currants, blueberries, and cranberries. Most of the tree fruit crops are grown in tropical and subtropical regions where the disease incidence is quite high and can be devastating to the commercial production.

Edible fruits are divided into two groups:

1. *Tree fruits*—Grow on woody trees (e.g., apples, pears, peaches/nectarines, apricots, plums, cherries, oranges, bananas, and pecans).
2. *Small fruits*—Grow on shrubs, vines, or herbaceous plants (e.g., grapes, currants, cranberries, blueberries, strawberries, blackberries, raspberries, kiwi, and haskap).

Table 16.1 illustrates the total world production of 8 top tree fruit crops. The world's top 6 small fruits are presented in Table 16.2.

Table 16.1   The world's top TREE FRUITS production per year (in million metric tons—MT) (FAO, 2019)

Rank	Tree fruit crop	Production (million MT/year)	Top countries
1	Apple	465.0	China
2	African oil palm	210.0	Malaysia
3	Banana	114.4	India
4	Mango	50.6	India
5	Orange	50.5	Brazil
6	Peach/nectarine	25.5	China
7	Lemon/lime	18.0	China/Mexico
8	Pear	7.5	China
	Total	**941.5**	

Table 16.2　The world's top SMALL FRUITS production per year (in million metric tons—MT) (FAO, 2019)

Rank	Small fruit crop	Production (million MT/year)	Top countries
1	Grape	74.3	Italy/China
2	Strawberry	5.8	USA/China
3	Currants (black, red, and white)	0.8	Europe (EU)
4	Cranberry	0.7	USA/Canada/Chile
5	Blueberry	0.6	USA/Canada/Poland
6	Gooseberry	0.2	Germany/Russia/ Poland
	Total	**82.4**	

16.3　Growing Fruits and Nuts

Not all fruit and nut crops have the same cultural requirements. It is important to recognize the climatic limitations and to understand the cultural management for each crop, especially if used in perennialization for ecosystem services. Listed below are some of those cultural and management considerations:

- *Site selection and slope*: Planting site requires a slight slope to encourage both air and water drainage—pockets of cold air can settle in low-lying areas and lead to frost damage and air circulation minimizes insect and disease problems.
- *Soils*: Deep soils encourage good root development, soils should be welldrained; crop preferences vary according to soil texture—examples:

  - Strawberries—sandy soils
  - Peaches—sand or silt loams
  - Cherries—silt loams
  - Pears—silt loams to clays
  - Apples—silt loams to clay loams

- *Water*: If natural rainfall is not sufficient, irrigation is needed to produce both quality and quantity of fruit. The water supplied must be both adequate in amount and of good quality.
- *Light*: Full Sun is necessary for high-quality fruit production, cloudiness, haziness, or air pollution can restrict light; proper pruning enhances light penetration into canopy for proper fruit development and ripening.
- *Temperature*: Length of the growing season must be adequate to produce the crop, Temperatures must be sufficient for optimum photosynthesis, sufficient cold weather is required to satisfy chilling requirement for certain crops.
- *Variety Selection*: Select varieties that perform well in specific climatic region! Consider: length

of growing season, chilling requirements, insect and disease resistance, soil adaptability, and fruit characteristics (size and color).

- *Pollination*: Many fruit varieties are *not self-fertile* and require *cross-pollination*, so another variety of the same species must be planted nearby for pollination, it must produce pollen at the same time, obtain this information from catalogs, extension agents, or nurserymen, check pollination charts for particular crops.

- *Labor*: For commercial production, labor supply is needed for operations such as planting, pruning, harvesting, packing, or processing.

- *Planting*: Fruit and nut trees are sold as bare-root, ball-and-burlap or containergrown, small fruits are usually sold as container-grown, proper spacing is important for optimum productivity.

- *Pruning*: In order to obtain efficient canopies with strong branches, and good light penetration, most fruit trees require some pruning. Effective pruning techniques promote efficient production of new growth, control the size of the tree, and provide for ease of harvest.

- *Thinning*: Fruit thinning decreases limb breakage by reducing the weight of the fruit load, remaining fruit develops to proper size, thinning may be done by hand, mechanically, or by chemical sprays.

- *Pest Control*: Monitor insects, if insect damage is unacceptable, a regular spray program may be needed, if birds eat or damage fruit, netting is available to cover fruit, remove fallen fruit and debris that could harbor insects and disease.

- *Weed management*: Chemical (herbicides), Organic (mulches + others).

- *Post Harvest*: Fruits consumed fresh contain large quantities of water—how does this affect perishability?

  – Which fruits can be preserved by drying?

  – Which fruits can be preserved by canning?

  – Freezing?

  – How perishable are nuts?

  – How have advances in transportation affected the availability of certain fruits?

  – What about refrigerated storage?

  – What about refrigerated transportation?

  – What about packaging?

- *Marketing*: This is perhaps the greatest challenge fruit crop growers face.

  – Identify Market and/or identify Market Niches:

    Product Differentiation and Certification

    Value-Added Processing

    Agri-tourism

    – Promotion Strategies.

    – Collaborative Marketing Strategies.

- *Consumer education*—a kcy rolc.
- *Health benefits* of consuming small fruit crops—the strongest marketing advantage.

    – Polyphenolic compounds, antioxidants, ellagic acid (HIV), vitamins; "one apple a day keeps doctor away".

- *Markets*: Home gardens, Local markets, worldwide markets, fresh and frozen products, dried fruit, juice, wine, canned and preserved fruit.
- *Value added products*: Juice, preserves, wine, etc.
- *Packaging and displaying* of fruits and fruit products.

16.4　Biotechnology and the Papaya: A Case Study in Hawaii

One example of an important tropical fruit with both excellent nutritional qualities and exceptional economic value that has been seriously threatened by a specific disease would be papaya in Hawaii (Fig. 16.1). The story of the survival of papaya in Hawaii shows the need for agricultural systems to embrace all aspects of scientific technology and not be blind to how we may merge principles of ecology with the wise application of biotechnology. The story begins in the 1940s, when papaya ringspot virus first made inroads into the papaya plantings in Hawaii. The disease causes distortion of leaves, mottling of fruit, and unmarketable appearance. This virus is spread from plant to plant by either mechanical means such as pruning or via aphid vectors. Several methods of control including netting and the use of insecticides proved to be ineffective and one by one the islands of Hawaii had to give up on the production of papaya. In some islands, production levels dropped by more than 90% and overall by the year 2000 overall production was less than 50% of previous yields and quality was greatly decreased.

The application of biotechnology techniques proved to be the ultimate sustainable, low-impact, and economic solution. A Hawaiian born scientist working at Cornell University was able to develop a genetically modified papaya, known as the Rainbow papaya with resistance to the virus. This has proved to be the salvation for the papaya industry in Hawaii. GM papaya now accounts for more than 90% of papaya production on the islands and has meant the resurgence of both large and small farms due to the approach of the introduction of the GM papaya. It was developed as a public–private partnership that provided seed throughout the islands. The use of GM technology has not displaced good stewardship but rather complemented it.

Fig. 16.1   Healthy papaya fruit (photo credit: K. Pruski)

16.5   Almonds and Water: Case Study

One of the most popular nut trees worldwide is the almond (Fig. 16.2). There is commercial production of almond trees worldwide in more than 40 countries on more than 2 million ha, an increase in acreage of more than 50% in the past 10 years. The almond is both a delicious and a nutritious food but it has one major drawback; it requires considerable water in order to be produced economically. This has meant that irrigation has been the norm in many of the major areas of production, such as the State of California in the USA. California is a major agricultural state which is dependent to a large extent on groundwater irrigation and almonds alone account for more than 10% of all of the state's annual agricultural water use. While almonds are an important money-maker for Californian farmers, they are not the only commodity with big demands on water. As the aquifers in California have been drawn down considerably over the last 50 years, the use of water by agricultural products has come into substantial competition. This applies not just to crops but also to the production of animal products such as milk.

Fig. 16.2　Almond orchard in Poland (photo credit: K. Pruski)

It is important in agroecology to assess aspects of production such as water use efficiency when making decisions on crops and agricultural systems. It has become obvious that California will need to make some very difficult decisions regarding water use in the near future, which will pit commodity against commodity and farmer against farmer. This same dilemma is facing many countries and producers around the world where water will become the defining factor regarding agricultural systems in the future.

For the moment, California's unique Mediterranean climate is almost ideal for almonds to flourish, but this is very much dependent on maintaining viable aquifers.

16.6　Agroforestry and Its Techniques

Agricultural systems that incorporate the use of woody perennials, such as fruit or nut trees, as well as more tropical products such as palms and bamboos, along with annual crops and even animal production, are collectively considered under the term agroforestry. Agroforestry systems are diverse and dynamic over space and time. Historically, small farmers have traditionally used basic agroforestry practices to diversify and stabilize their production and income. Agroforestry systems are classic multifunctional units that are meant to provide diverse products and maintain a sustainable (economically and environmentally) production.

Since agroforestry systems have historically been developed to respond to the economic needs of individual farmers in individual locations, they are as diverse as the farmers and the location in which they were developed. However, the basic idea in any agroforestry system is that of encouraging positive interactions on the ecological side and diversifying opportunities for products for sale or use by the farmer. Mixtures of annual and perennial crops in some type of intercropping strategy may be separated spatially or temporally. In general, short-term returns come from annual crops whereas longer-term returns are the result of perennial crops and livestock. In addition, there is an attempt to close nutrient cycles by utilizing complementary rooting structures and types of crops (e.g., legumes with grasses) and the use of animal manures to stimulate plant growth and increase soil sustainability. Many different systems are in use around the world today, and as the interaction of trees and food crops is better understood, more will be developed.

Although there is a wide diversity of agroforestry systems, it is possible to classify them into four basic groups, depending on the types of crops and/or livestock utilized:

- Agri–silvicultural systems: This is an intercropping system in which the farm income comes from both food and wood products.
- Silvo–pastoral systems: This is a system in which trees are mixed with pasture for livestock and therefore the income is from wood products and livestock.
- Agri–silvo–pastoral systems: As the term suggests, this is a more complex system in which there must be a balance among food crops, space for pasture and trees for wood products. The management of this type of system changes as the ecosystem matures and the trees become larger. However, the level of diversity and stability both from an environmental and economic standpoint is high.
- Mixed garden systems: This is a particular type of agri–silvo–pastoral system that is a true adaptation of a particular farmer to particular land base and economic need. It integrates trees, crops, and animals on small plots with the overall purpose very much designed for the supply of nutrients, materials, and marketable products for a family. Any one such particular system is not readily transferable to another farm.

Agroforestry illustrates the need for using agroecology concepts and principles to integrate diverse natural components of ecosystems by taking information from both traditional farming and modern agricultural technology to make positive change. The diverse disciplines inherent in understanding and building agroforestry requires both coordination and communication which is the key to agroecology.

Conclusion

The incorporation of perennial crops into agricultural systems has huge potential for positive impact both environmentally and economically. However, there is a need for more research to quantify the effects and to find the correct balance between environment and economics both in the short term and in the long term. This will involve appraising both simple and complex approaches to incorporation of diversity in perennial/annual/livestock systems. We need to consider carefully both principles and practice as they apply in both small landholders and larger farm systems.

Chapter 17
Aquaculture Production Systems

J. Duston and Q. Liu

> *"We must plant the sea and herd its animals... using the sea as farmers instead of hunters. That is what civilization is all about—farming replacing hunting."*
> —*Jacques-Yves Cousteau*

Abstract The global wild fishery cannot provide for the increasing demand for fish, molluscs, crustaceans and aquatic plants. Aquaculture is the farming of these aquatic organisms that is increasing to meet this demand. The dramatic increase in aquaculture to provide nutrition for growing populations is also increasing stress on the environment. There is need to develop sustainable, energy-efficient methods in the industry. This chapter examines techniques for how we can have a sustainable aquaculture globally.

Learning Objectives

After studying this topic, students should be able to:

1. Explain why aquaculture must grow in the twenty-first century.
2. Describe the role of aquaculture in human nutrition.
3. Outline and describe methods for energy efficiency in finfish aquaculture.
4. Discuss the prospect for sustainability of aquaculture from a global perspective.

17.1 Aquaculture Must Grow in the Twenty-first Century

Aquaculture is the farming of aquatic organisms, including fish, molluscs, crustaceans and aquatic plants. Since 1970, aquaculture production has increased severalfold to meet the demand resulting from the growing population and declining wild fish stocks (Fig.17.1; Pauly and Zeller, 2016; FAO, 2020). However, the popular claim that aquaculture is the fastest growing food sector in the world over the past few decades has been challenged recently; it seems poultry production is expanding fastest (Edwards et al., 2019). The aquaculture industry's contribution to the total food supply has increased at a rate of 10%–11% per year in 1980s and 1990s, and 5.8% between 2000 and 2016 (Fig. 17.1; FAO, 2018). Compared to <1 million tons annual production in 1950s, world aquaculture in 2018 was 82.1 million tons (MT) of food fish, and 32.4 MT of aquatic plants, accounting for about 46% of the total seafood products (Fig. 17.1; FAO, 2020). That world aquaculture production exceeded beef production in 2012 was reported widely (e.g. Béné et al., 2015), but a reassessment puts it about 60% of beef production (Edwards et al., 2019). Nevertheless, the lower food conversion ratio (FCR: the feed requirement in kg per kg body weight gain) values of aquaculture species [e.g. salmon (0.95–1.5) : 1; shrimp (1.7–2) : 1] and various terrestrial animals [e.g. cattle (5.15–6.95) : 1; poultry (2.13–2.61) : 1] are not only important for reducing the production costs but also have less environmental burden to bear (Naylor et al., 2009; Béné et al., 2015). Moreover, the availability of terrestrial space for agriculture and livestock farming is becoming increasingly limited. Much of the world's grassland is stocked at or beyond capacity. Our world is comprised of 70% water and most of it, the oceans, is under-utilized due to inadequate knowledge and resources.

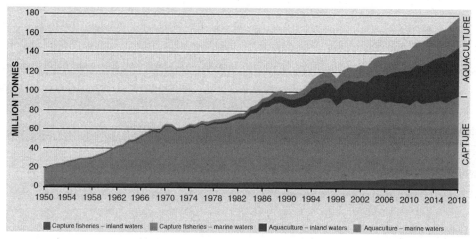

Fig. 17.1 World capture fisheries and aquaculture production between 1950 and 2018 (FAO, 2020)

Excludes aquatic mammals, crocodiles, alligators and caimans, seaweeds and other aquatic plants

The development of aquaculture involves the domestication of an increasing number of aquatic organisms derived from the wild (>400 spp.) and intervention of their culturing environment including both inland and marine areas (Teletchea and Fontaine, 2012). Freshwater fish represent about 62.5% of total world aquaculture production in 2018, dominated by carp, catfish and tilapia (FAO, 2020). The expansion of marine species such as shrimps, salmon and bivalves is also evident in recent years due to the strong market demand. China plays a large role in the development of aquaculture. It is the leading producer, exporter, processer and consumer of farmed finfish and shellfish as of 2018, producing 58% of global aquaculture volume but also consuming 38% of the global production with per capita consumption reaching about 41 kg, driven by a rapidly expanding aquaculture sector and growing middle-class population (Cao et al., 2015; FAO, 2020).

17.2   The Role of Aquaculture in Human Nutrition

Farmed fish and aquatic plants represent a valuable source of both macronutrients and micronutrients of fundamental importance for a healthy diet. Fish can be an essential component of a nutritious diet, as high levels of long-chain poly-unsaturated fatty acids contribute to neurodevelopment during the most crucial stages of an unborn or young child's growth (Swanson et al., 2012). Long-chain omega-3 fatty acids, including eicosapentaenoic acid (EPA) and docosahexaenoic acid (DHA), are mainly found among cold-water fish, shellfish and aquatic algae. Intake of fish or long-chain omega-3 fatty acids supplements that contain EPA and DHA have been recommended by maternal nutrition guidelines, since human bodies do not efficiently produce them. In addition, fish consumption may benefit mental health and prevent cardiovascular disease, stroke and age-related macular degeneration (Swanson et al., 2012). In developing countries in Africa, fish consumption is about 10 kg in 2015, much lower than the world average of 20 kg (Obiero et al., 2019). Micronutrient deficiencies are related to an estimated one million premature deaths annually (Hicks et al., 2019). In Sub-Saharan Africa alone, the number of malnourished people is about 224 million (Obiero et al., 2019). The problem of micronutrient deficiencies is further exacerbated by unbalanced fish supply, mainly by fisheries on coastal regions, and the rapid growth in population, projected to double by 2050 (Obiero et al., 2019). The limited fish supply, however, represents an important source of animal protein and micronutrients. While average per capita fish consumption may be low, even small quantities of fish can provide essential amino acids, fats and micronutrients, such as iron, iodine, zinc, vitamin D and calcium, which are often lacking in vegetable-based diets. To date, aquaculture is only practiced among a few countries and accounts for about 16% of total fish production in Africa (1.6 out of 10.4 million tons; Obiero et al., 2019). Considering that most of the wild fish populations are fully exploited in Africa, fish farming has great potential to meet the growing demand of fish and will be an important means to improve the per capita fish consumption and combat the problems of malnutrition and food insecurity (Hicks et al., 2019; Obiero et al., 2019).

Aquaculture is the main source of aquatic plants, accounting for 96% of production in 2016 (FAO, 2018). Seaweeds and other algae are an important part of diets in several cultures, particularly in Asia. The most widely cultivated species intended for human food include Japanese kelp (*Saccharina japonica*) and wakame (*Undaria pinnatifida*), which are popular for use in soups, and the red seaweed nori (*Pyropia and Porphyra* species) is used to wrap sushi. The nutritional contribution of seaweeds consists mainly of micronutrient minerals (e.g. iron, calcium, iodine, potassium, selenium) and vitamins, particularly A, C and B-12. Seaweed is also one of the only non-fish sources of natural omega-3 long-chain fatty acids (Fig. 17.2).

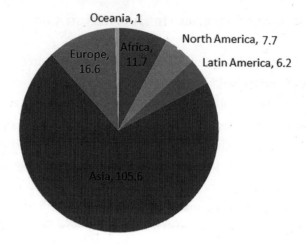

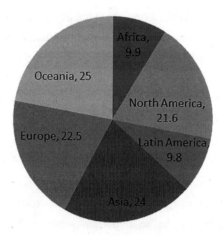

Fig. 17.2　Total (million tons) and per capita (kg/year) food fish consumption by region: Africa, North America, Latin America, Asia, Europe and Oceania, in 2015 (FAO, 2018)

17.3　Methods and Energy Efficiency in Finfish Aquaculture

17.3.1　Freshwater Ponds and Lakes

Ponds, constructed holes in the ground filled with freshwater, are the oldest form of aquaculture. The earliest evidence of fish farming dates over 3000 years ago. In the Zhou dynasty (1112BC–AD221), Fan Li, around 500 BCE, described carp farming. In Europe in the Middle Ages, the monastic orders and the aristocracy were the main users of freshwater ponds, since they had a monopoly over the land, forests and water courses. Traditionally, growing fish in ponds was a good example of extensive aquaculture, the fish fed solely on natural food, the farmer perhaps providing fertilizer to promote phytoplankton growth at the base of the food chain, which in turn supported growth of zooplankton and larger predators. In its most refined form, China developed polyculture aquaculture, where various species of cyprinids are reared together, each with specific feeding requirements. Silver carp (*Hypophthalmichthys molitrix*) feed on microalgae and microzooplankton, grass carp (*Ctenopharyngodon idella*) feed on macrophytes, bighead carp (*Aristichthys nobilis*) and common carp (*Cyprinus carpio*) both feeding on zoobenthos and microzooplankton. Ponds filled with freshwater are still the main form of aquaculture, contributing 60% of world production in 2008 despite freshwater contributing only 3% to global water resources (Bostock et al., 2010). Cyprinids remain the most important species grown in China, but production of tilapia and catfish is also very large in countries closer to the equator where temperature is high most of the year. Tilapia are mostly *Oreochromis niloticus*, derived from the River Nile, Egypt. Striped catfish (*Pangasianodon hypophthalmus*) is grown in Vietnam and exported around the world, putting pressure on US growers of channel catfish (*Ictalurus punctatus*). Modern pond production is now either semi-intensive or intensive, meaning either some or all of the food for the fish is a formulated pelleted diet.

Pond depth is typically about 1.5–2 m, with a flat bottom to allow the fish to be captured efficiently by seine net. The recommended area of a catfish pond in the USA is 4–6 ha. Advantages of ponds: low operating costs, good for specific species that tolerate high temperature, high turbidity, but generally unsuitable for salmonids as they prefer clear running water. Disadvantages of ponds: no biosecurity, fish can be exposed to harmful pathogens and predators (e.g. birds). At harvest, capturing the fish in large ponds is difficult. Hypoxia is a big concern at high temperatures when availability of O_2 is low and fish O_2 demand is high. Risk of mortality is greatest at night since photosynthesis ceases and plants consume oxygen. During daytime substantial amounts of oxygen are produced by photosynthetic plants (micro- and macro-algae). Organic sediment from fish waste on the bottom of the pond also consumes O_2 and must be removed after harvest by first draining the pond then using mechanical diggers. Oxygen supplementation of ponds, often by paddle-wheels aerators can greatly increase production efficiency, but is not always an option in remote locations lacking reliable power supply.

17.3.2   Cages in Fresh- and Sea-water

Cages in freshwater, also referred to as net-pens, allow farmers to manage their fish more carefully and allow deep lakes to be utilized. In Canada, for example, rainbow trout (*Oncorhynchus mykiss*) is grown in cages in Lake Huron, Ontario and Bras d'Or Lake in Nova Scotia. The net-pens, up to 6 m deep, are suspended from plastic floating circles up to 100 m circumference, providing a large rearing volume of about 4750 m^3. All the food for the trout is provided by the farmer in the form of pellets, defining it as intensive aquaculture. Underwater cameras ensure the food is consumed and not lost through the bottom of the cage. Within the cage, the fish are protected from predators, and can be easily harvested. Oxygen is the principal limiting factor to production due to the poor water exchange through the cage. Consequently, stocking density of rainbow trout cannot exceed 20 kg/m^3 of water, and feeding must be strictly monitored since the digestion process greatly increases oxygen demand. Tilapia, by contrast, are tolerant to hypoxia, and are safely reared in large numbers of small cages <10 m^3, sharing the same pond.

Cages in tidal coastal regions have the great advantage of good water exchange providing a good supply of oxygen. In Canada, coastal British Columbia, New Brunswick, Newfoundland and Labrador and Nova Scotia have large numbers of floating cage systems with an annual production of about 120,000 tons of Atlantic salmon (*Salmo salar*; Statistics Canada). The temperature range of 2–15℃ is ideal for Atlantic salmon, but in recent years temperatures >15℃ are becoming more common, posing a threat to the industry. The world leader in Atlantic salmon farming is Norway, producing 1.3×10^6 tons in 2018 (Statistics Norway). A serious disadvantage of sea-cages is the escape of Atlantic salmon due to net breakage due to storms. Aside from the loss of money to the farmer, there is accumulating evidence of escaped farmed fish inter-breeding with wild Atlantic salmon, causing loss of fitness due to genetic introgression (Glover et al., 2017). Moreover, concerns that pathogens from farmed salmon are exiting the cages and threatening wild stocks are leading to increased political pressure to close coastal sea-cage operations in British Columbia (Morton and Routledge, 2016).

17.3.3　Land-Based Aquaculture

In Europe and North America for over 100 years, land-based aquaculture, in which fish are held in tanks rather than ponds, was primarily for rearing salmon for restocking into rivers and rainbow trout to "pan-size" for food. The development of this technology was the foundation on which the industrial production of salmon was established by Norway in the 1970s. The salmon 'revolution' was to add marine grow-out in net-pens to the long established techniques for rearing from egg to smolt in freshwater on land. The persistent criticism that marine net-pens are not ecologically sustainable is driving investment into wholly land-based production of salmon, the economic sustainability of which is uncertain. Summarized below is land-based aquaculture of Atlantic salmon, largely from a Canadian perspective, but considered broadly applicable internationally.

The traditional salmon hatchery has two sources of freshwater, groundwater and surface water gravity fed from either a lake or river. Groundwater in Canada is about constant 9℃ year-round, providing a good temperature for the egg and post-hatch yolk-sac stages. At "swim-up" the young salmon, about 2.5 cm long, begin feeding on a fine pelleted diet. Heating the water within the indoor hatchery to >14℃ is necessary to improve the growth-rate of the young salmon during the cold Canada spring. The high financial cost of heating drove the development of water recirculation systems, where the water was filtered and recycled. Salmon were reared indoors in recirculation systems until June when the water in the outdoor tanks (3 m diameter) was warm enough (>15℃) to maintain rapid growth. By May the following year, the juvenile salmon transform from parr to smolt (50–100 g), a special physiological state that allows the fish to be directly transferred from freshwater to seawater net-pens.

Modern salmon hatcheries in Canada are entirely indoors and wholly reliant on recirculation technology to maintain optimum temperature and water quality from the egg through to smolt stage. The fish are reared in large tanks (10 m diameter). The settleable faeces and uneaten food are removed by swirl separators, then a 30 µM drum filter removes suspended sediments. Toxic ammonia excreted by the fish is converted by nitrifying bacteria (Nitrosomonas, Nitrobacter species) in huge biofilters to less toxic nitrite (NO_2^-) and relatively safe nitrate (NO_3^-). Then, toxic carbon dioxide (CO_2) excreted by the fish and the biofilter is removed by a degasser tower and vented off outdoors. The final step of the recirculation process is the injection of pure oxygen (O_2) into the water which is then returned to the fish tanks. Precise control of temperature and pH is achieved by automated control systems, resulting in rapid and efficient smolt production. The maximum body weight of smolts has increased progressively from 60 g in the 1990s to as high at 300 g today. The advantage of larger smolts is a shorter grow-out period in the marine net-pens by several months. A shorter marine phase reduces the time the salmon are exposed to the risks of sea-lice, and extreme weather events that can cause net failure and escape. Sea-lice have become a crisis since the ectoparasite acquired resistance to a once highly effective chemical therapeutant (Aaen et al., 2015).

Full land-based production of Atlantic salmon from egg to 5 kg market size using state-of-the-art recirculation technology is technically feasible and is being conducted around the world, including Nova Scotia (Sustainableblue.com). Since "100% land-based" eliminates the controversial sea-cages, it enjoys full support of environmentalists and government agencies, yet the economic sustainability of such systems is unclear. Moreover, the carbon footprint of a land-based system is twofold higher than a comparable sea-cage system (Ayer and Tyedmers, 2009; Liu et al., 2016). For example, when 52% of the electricity in Nova Scotia in 2018 came from burning coal and petcoke (NSPower.ca), the "sustainability" of land-based salmon farms with a large power demand for pumping water is unclear. In China, water recirculation pumps consumed an estimated 37% of electricity costs of 7500 kWh per ton of salmon grown from smolt to 5 kg in a RAS in China (Song et al., 2019). 145 tons of salmon were harvested from a farm with 6000 m³ of tank space (Song et al., 2019). Life-cycle assessment (LCA) approach has been used to estimate the sustainability of land-based salmon farming (Liu et al., 2016; Song et al., 2019). The estimated carbon footprint of land-based salmon production was twofold higher than the traditional sea-cage model (Liu et al., 2016).

17.4　Sustainability of Aquaculture: Global Perspective

The rapid development of aquaculture over the past 40 years has been celebrated as the "Blue Revolution". A fundamental limiting factor that may inhibit further expansion is the reliance on fishmeal and fish oil derived from the capture fishery as the principal ingredients for the food pellets fed to farmed fish (FAO, 2018). In 2011, 23 million tons of small pelagic marine fish such as anchovy, herring, mackerel and sardines were used for both meal and oil for farmed animal feeds, most of which was used for aquaculture (73%), followed by pigs (20%), poultry (5%) and others (2%; Shepherd and Jackson, 2013). These small pelagic fish are vital to the marine food-web supporting larger fish, birds and marine mammals. Overexploitation, poor management and climate fluctuations can all threaten the abundance of these wild fish populations. In Chile, there is evidence that the wild fishery is being overexploited, and that the large salmon farming industry there is adding to the demand for fish-meal and -oil. This leads to the unsettling conclusion that the Blue Revolution may be going in the wrong direction (Nahuelhual et al., 2019).

Concern over the overuse of wild fish in aqua-feeds is greatest for carnivorous species, such as Atlantic salmon. An estimated 4.5 kg of wild fish is needed to produce 1 kg of Atlantic salmon, compared to 0.2–1.4 kg for 1 kg of omnivorous and herbivorous fish such as carp and tilapia (Naylor et al., 2009). In China, the culture of carp and tilapia in freshwater ponds dominate, accounting for 90% and 50% of global production (Cao et al., 2015; FAO, 2018). Fishmeal inclusion rates in feeds are low, about 6% for tilapia and 3.2% for carp, whereas fish oil inclusion is minimal (Cao et al., 2015). Nevertheless, the massive scale of carp and tilapia production in China demands fish meal imports of about 10 million tons per year, about fourfold greater than the entire European Union (Guillen et al., 2019). On an optimistic note, the dietary inclusion rates of fishmeal and fish oil in aqua-feeds for salmonids are declining, being partially replaced by crops such as oilseeds and livestock by-products (Colombo et al., 2018). Fishmeal and fish oil inclusion rates in Atlantic salmon diets between 1990 and 2013 decreased from 65% to 18% and from 24% to 11%, respectively (Ytrestøyl et al., 2015). A recent study indicated that the dietary inclusion rates of both ingredients can be reduced to about 5% without affecting somatic growth in Atlantic salmon (Foroutani et al., 2018). In addition, the food conversion efficiency among all farmed fish is improving because of better feed formulations, manufacturing methods and feeding practices (Naylor et al., 2009; FAO, 2018).

Waste products from farmed fish released into the environment pose a potential threat to aquatic ecosystems. Settleable solids create anoxic mats, and dissolved nitrogen and phosphorus promote unwanted algae growth and marine sea-cages in coastal regions. Deposition of organic waste beneath sea-cages has, for many years, been identified as a threat to the extremely valuable lobster fishery. The contradictory evidence illustrates the complexities: one study concluded salmon farms caused significant harm, another reported no problems (Milewski et al., 2018; Grant et al., 2019).

Concerns over anthropogenic eutrophication stimulated much research into "integrated multi-trophic aquaculture" (IMTA), combining fish and other aquatic animals and plants to remove particulate and dissolved wastes from fish farming, and thereby producing a self-sustaining source of food (Neori et al., 2004). The commercial viability of IMTA remains to be proven, but one day perhaps the integration of species will attain the same efficiency as the ancient Chinese art of carp farming.

Chapter 18
Environmental Innovations in Urban Ecosystems

T. McKenzie

"Cities have the capability of providing something for everybody, only because, and only when, they are created by everybody."
—*Jane Jacobs*

Abstract As the global population becomes more urbanized, cities continue to expand and consume valuable agricultural and forested land as well as biologically diverse wetlands. Urban development degrades natural ecosystems through fragmentation, pollution, the introduction of barriers, exposure, and destruction of natural elements. These disturbances cause changes to wildlife/species composition and distribution resulting in a significantly altered ecological system that is not only influenced by humans but also specifically designed and created by them. This urban ecosystem is dominated by various forms of infrastructure that provide services for humans including green infrastructure that is made up of both natural vegetative systems and green technologies. Green infrastructure provides a multitude of provisional, regulating, supporting, and cultural ecosystem services. This includes, but is not limited to, agricultural production, pollution abatement, and storm water mitigation. The value of these services to the inhabitants of the city justifies more consideration of green infrastructure during the planning, design, construction, and maintenance of urban spaces. Geographic information systems (GIS) provide an excellent mechanism for assessing urban growth and determining the extent of important green infrastructure components and their associated ecosystem services provided.

Learning Objectives

After studying this topic student should be able to:

1. Define the following terms:

- Urbanization
- Sprawl
- Fragmentation
- Succession
- Green infrastructure
- Peri-urban forest

- Wildland–urban interface
- Agro-park
- GIS
- Extensive and intensive green roof

2. Describe the impacts of urbanization and "sprawl" on natural ecosystems.

3. Describe how urban forests, green rooftops, storm water systems, and urban parks contribute to green infrastructure.

4. List and briefly describe environmental services provided by plants growing in urban environments.

5. List three examples of storm water management strategies.

6. Explain the role of GIS in green infrastructure planning.

18.1　Urbanization Impacts

18.1.1　What Is Urbanization?

Increased urbanization is a present global trend; greater proportion of people are living in high density areas. The United Nations predicts that this trend will continue from its present value of about 55% of the world's population living in urban areas to a level of 68% by 2050 (United Nations, 2018). Unprecedented economic growth in the western world has encouraged population growth and a trend toward urban. In 2018, 82% of the people in North America lived in urban areas, closely followed in density by Latin America and the Caribbean (81%), Europe (74%), and Oceania (68%).

These changes in population, land-use, and development patterns are increasing the space requirements for urban centers. Suburban population of cities throughout the world continues to expand at a rate which is often greater than what is seen at the urban core. The land most often consumed by suburban development is agricultural land or natural areas. For the latter, development can lead to a reduction in forested land in and around suburban areas and have negative impacts on remaining natural ecosystems. Natural ecosystems can be significantly altered by human activity through the introduction of new organisms and disturbances in the ecosystem as a result of development.

18.1.2　Sprawl

Sprawl is defined as dispersed and inefficient development that is characteristically low density, automobile-dependant, and environmentally damaging (Hasse and Lathrop, 2003). Sprawl has been described as haphazard, creating a patchwork of low-density and commercial strip development that fragments, degrades, and isolates remaining natural areas with significantly higher economic and social costs than more compact developments. Unfortunately, many municipal planning and development programs fail to recognize these conditions as problematic or threatening to our environment.

The impact of development on wildlife habitat as well as on water and air quality has been well documented (Friesen, 1998; McPherson, 1998). Current planning practices in most municipalities do not identify sensitive environmental areas and natural ecosystems or recognize the impact development has on the environment. More emphasis is usually placed on zoning restrictions for what type of development is appropriate or permitted in a particular area. As a result, conservation of natural areas and allowances for greenspace are often overlooked during the city planning process (Mahon and Miller, 2003).

Scientific research has also identified some of the impacts of urban development on natural ecosystems including chemical, physical, and biological change (Nowak, 1994c), exposure and destruction of natural elements (Nichols, 1999), fragmentation (Collinge, 1996), the creation of barriers (Nichols, 1999), and changes to wildlife species composition and distribution (Friesen, 1998). Chemical stress occurs in the ecosystem as a result of the introduction of common urban pollutants such as carbon monoxide, nitrogen oxides, sulfur dioxide, ozone and airborne particulate matter from urban soils, industrial processes, combustion, and chemical reactions in the atmosphere (Nowak, 1994a). Development results in physical destruction of ecosystem components, the replacement of natural conditions with development infrastructure such as buildings and roads, and the introduction of non-native species resulting in a change in species diversity. Soil conditions are usually modified and impervious surfaces are introduced resulting in changes to surface drainage patterns and increases in site temperature due to reflection of heat by paved surfaces (Akbari et al., 1992).

18.1.3   Fragmentation

As cities and suburban areas expand into undeveloped forested areas and agricultural land, development and associated infrastructure in the form of roads and utility corridors fragment forest and agricultural land. Forest fragmentation is defined by the USDA Forest Service (Rowantree et al., 1994) as the division of forests into smaller separate units by roads, utility rights-of-way, and other developments. It decreases interior forest habitats for animals, and introduces pollutants, non-native species of plants and animals, and domestic animals that may harm native plants and animals. The fragmented forest continues to change over time due to other anthropogenic factors. Natural fragments tend to disappear as development progresses and vegetative cover patterns change with the addition of human-made landscapes of turf, exotic plant species, street trees, and paved surfaces. Fragmentation of ecosystems by developmental infrastructure also increases the amount of edge or border that exists between developed land and natural habitat. "The changes in light, moisture and wind, most pronounced at the fragment edge, may significantly alter the plant and animal communities which occur there" (Collinge, 1996).

　　Barriers to movement include roads, open areas, buildings, and other paved surfaces such as parking lots. Habitat is broken into smaller segments with varying levels of connectivity between segments. The number of plant and animal species is often influenced dramatically by the placement of barriers. Many naturally occurring plants are relatively immobile and ill adapted for re-colonization following human disturbance (Friesen, 1998). As areas become urbanized, the diversity of indigenous or natural species tends to decline, but quantity of some species increases as a result of their ability to adapt to urban conditions. New species are introduced to the urban environment during landscape development. The result is a species composition and/or community that is radically different than that of the natural habitat which preceded development (Turner et al., 2004).

18.1.4　Forest Disturbances and Change

　　Change in species composition and ecosystem condition is the norm in natural forests. "Often when a forest is severely disturbed by natural forces or by complete removal of trees through logging, the new forest is composed of different species compared to the original species" (Kotar, 2000). Changes in the species composition of an ecosystem over time often occur in a predictable order. In forests, this is referred to as succession, which is the sequence of one community of plants gradually replacing another (Canadian Forest Service, 1995). Suburban development on existing agricultural land and natural areas can cause both micro disturbances and community-replacing disturbances which can have an impact on succession of a natural ecosystem. Suburban development has been found to introduce more frequent acute disturbances, as well as chronic disturbances (Rowntree, 1988).

18.1.5　Disturbances Related to Urban Development

　　As development extends into natural areas, fragmented forest patches are exposed to four general categories of forces that may lead to a disturbance of the existing ecosystem (Fig. 18.1). These categories of forces have been described as direct human, indirect human, direct natural, and indirect natural (Nowak, 1993).

　　The frequency and intensity of disturbances as a result of human activity can create an unnaturally dynamic environment causing severe changes in an ecosystem. Such rapidly changing conditions often result in significantly variable vegetation patterns and species composition across the urban landscape (Dwyer et al., 1999).

18.2   Is the City an Ecological System?

18.2.1   Cities as Ecosystems

The recognition of the city as an ecological system has been widely debated in the scientific community. The city obviously contains a collection of living and non-living things interacting within a defined boundary. Many feel the type of interactions that occur in the city is far from natural and heavily influenced by humans as one dominant organism. It has also been suggested that humans have become a dominant influence on almost all of the earth's ecological systems (Vitousek et al., 1997).

The term *urban ecosystem* was introduced in the early 1970s as an ecological system that was not only influenced by humans but more specifically designed and created by them (Stearns and Montag, 1974). These new synthetic ecosystems were recognized as being significantly different from other human influenced natural ecosystems due to design, development, sphere of influence, and potential impacts. The study of these ecosystems has been recognized as an interdisciplinary field influenced by both social and natural sciences (Mcintyre et al., 2000).

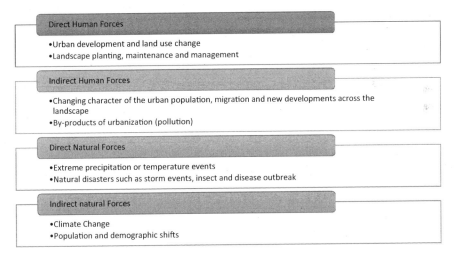

Fig. 18.1   Examples of forces that may disturb natural ecosystems

18.2.2   Infrastructure and the Built Environment

Humans influence both the living and non-living components of the urban ecosystem. The non-living component is usually dominated by human infrastructure (Fig. 18.2). Infrastructure is a widely accepted term for the human built environment consisting of buildings and places as well as the physical connections between places that carry people, materials, information, and energy.

These "fixed" things include roads, railways, pipes, and cables and are often collectively referred to as *hard infrastructure*.

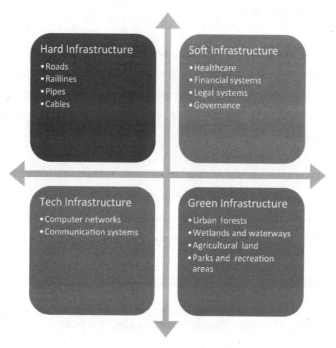

Fig. 18.2   Urban infrastructure

An urban area also contains social and economic infrastructure. Humans and their systems of interaction and organization such as healthcare, finance, laws and governance that provide a critical service to the people that live there define levels of interaction and are commonly referred to as *soft infrastructure* (Gu, 2017). Two other forms of urban infrastructure are beginning to receive more attention for the unique services they provide. *Tech infrastructure* or information technology (IT) infrastructure is rapidly expanding the range of services available to improve interactions and resource management with an urban setting. Increasingly sophisticated tools and systems make us smarter and improve the way we build communities and connect people.

Green infrastructure is defined as the natural vegetative systems and green technologies that collectively provide society with a multitude of economic, environmental, and social benefits (Green infrastructure ontario, 2019). Common forms of green infrastructure include:

1. Urban forests and woodlots.

2. Wetlands, ravines, waterways, and riparian zones.

3. Storm water systems.

4. Agricultural lands/urban agriculture and meadows.

5. Green roofs and green walls.

6. Parks, gardens, and recreation areas.

18.2.3   The Urban Forest

Nowak (1994a) described urban forests as complex ecosystems created by the interaction of anthropogenic and natural factors. Urban forest ecosystems may occur within an urban ecosystem which is composed of the various plants and animals that once existed in former agricultural or natural areas with the addition of other organisms including humans and human infrastructure. Urban forests are an integral part of this urban ecosystem and have been described as including the sum of all woody and associated vegetation in and around dense human settlements, ranging from small communities in rural settings to metropolitan regions (Miller, 1997).

18.3   Urban Ecosystem Services

18.3.1   Ecological Services Provided by the Urban Forest

An important role of the urban forest is the provision of ecological services. Gretchen Daily has defined ecological/ecosystem services as "the conditions and processes through which natural ecosystems, and the species that make them up, sustain and fulfil human life" (Daily, 1997). The Millennium Ecosystem Assessment (2005) categorized these services into provisional, regulating, supporting, or cultural services (MEA, 2005). Research has identified the following important ecological services provided by the urban forest: energy conservation (McPherson, 1994); carbon storage and sequestration (Nowak, 1994b); removal of gaseous pollutants and particulate matter from the air (Nowak, 1994a); storm water mitigation (Miller, 1997); and wildlife habitat (Nichols, 1999; Dunster, 1998). Studies have also revealed that the body of knowledge regarding the role of vegetation in the urban ecosystem and for enhancing human well-being is inadequate for managers to make informed decisions (Rowantree et al., 1994) (Fig. 18.3).

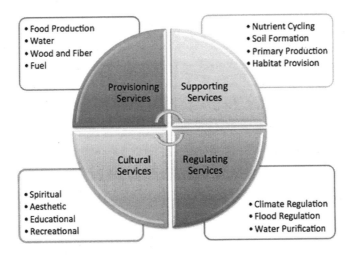

Fig. 18.3   Ecosystem services (Millennium Ecosystem Assessment Program, 2005)

18.3.2　Energy Conservation

Trees play an essential role in urban environments by moderating local temperature. "Trees and vegetation affect urban meso- and microclimates on three levels: human comfort, building energy budgets, and urban mesoclimate" (Miller, 1997). Mesoclimate refers to minor variation of general climatic conditions in a region due to effects caused by topographic features, bodies of water, and other influences. Trees influence local climate by affecting solar radiation, air movement, air temperature, and humidity.

Tree canopies intercept ultraviolet radiation of the sun, thus reducing the amount of energy reaching other urban surfaces. The shade provided by vegetative canopies results in a decrease in the amount of heat absorbed, stored, and radiated by built surfaces. Studies have shown that temperatures within greenspace may be 3℃ lower than outside the greenspace (McPherson and Rowantree, 1993). Individual trees have been found to provide cooling under their canopy of 0.7–1.3℃ (Souch and Souch, 1993). Canopies of trees also provide a barrier and/or alter wind movement in urban environments. This can assist in cooling by channeling summer breeze or provide shelter from cold air. It has been found that scattered trees throughout neighborhoods can reduce ground-level wind speeds by as much as 50% (Heisler et al., 1994). Evapotranspiration (ET) is a natural process associated with tree growth that occurs when water is lost by transpiration from the leaf surface and evaporated from the soil under the tree canopy (Harris et al., 1999). ET is influenced by sunlight, temperature, humidity, and wind and has been found through simulation studies to have a considerable cooling effect (McPherson, 1994).

The cumulative impact of trees on solar radiation, air movement, and relative humidity can have a significant impact on energy use in buildings by reducing the need for energy to provide heating and cooling. Trees reduce potential heat loss from buildings by reducing air infiltration, heat conduction, and radiation transmission. Windbreaks formed by trees have been found to reduce home-heating costs by 4%–22% (Miller, 1997). Properly placed trees can also provide shade for buildings and influence air movement and humidity around and within structures to reduce the need for summer cooling. A single 7.5 m tree is estimated to reduce annual heating and cooling costs by 2%–4% if located properly (McPherson, 1994). These savings, if applied to an entire urban region, can have significant impact. A study in Sacramento, CA, suggests that the existing urban forest is responsible for annual air conditioning savings of approximately $18.5 million per year of electricity or 12% of total air conditioning costs and 1.5% of electrical use for the county (Simpson, 1998).

18.3.3　Carbon Storage and Sequestration

Uptake of carbon dioxide (CO_2) during photosynthesis enables trees to assist in the reduction of

CO_2 in the atmosphere. Carbon dioxide is stored in the accumulation of woody biomass as trees grow. The storage of CO_2 in above- and below-ground biomass has been referred to as carbon sequestration and is commonly expressed as a rate of carbon stored per year (McPherson, 1998).

Carbon storage by urban forests in the United States has been estimated at 350–750 million tons (Rowantree and Nowak, 1991). Several North American cities have been assessed for potential carbon sequestration of their urban forests. Trees in Chicago, IL, store 85.7 tons per ha in comparison to trees in Oakland, CA, at 40 tons per ha. Differences in amounts stored are due to differences in urban forest structure such as tree cover, species, and growth rate (Nowak, 1994b). The capacity of trees to grow and store biomass obviously limits their temporary carbon storage potential as tree death ultimately leads to release of CO_2 through decay or fire.

Society has recognized the importance of temporary carbon storage in plant biomass and the need to offset losses in forest biomass with tree planting programs. The Tree Canada Foundation was established in 1992 in response to concern regarding climate change. The objective of the Tree Canada Foundation is to encourage Canadians to plant and care for trees in an effort to help reduce the harmful effects of carbon dioxide emissions. This program has assisted in planting 170 million trees across Canada and hopes to encourage the planting of another 90 million. Achieving this goal would establish enough trees to sequester 5.2 million tons of CO_2 in 10 years at an average of 0.8 kg per tree per year (David, 1996).

18.3.4 Removal of Gaseous Pollutants and Particulates

Gaseous pollutants and particulate matter in the atmosphere are primarily the result of burning fossil fuels, industrial processes, soil erosion, and reactions between sunlight and pollutants (Harris et al., 1999). Trees function in pollution reduction by absorbing particles and gases on their leaf surfaces. Air pollutants are removed from the air primarily by three mechanisms: wet deposition, chemical reactions, and dry deposition (Fowler in Nowak, 1994a). Deposition to plant material occurs in three ways: sedimentation under the influence of gravity, impaction by wind, and deposition by precipitation. Gaseous pollutants are predominantly dry-deposited through leaf stomata with less deposition occurring on the leaf surface (Smith, 1978). The main gaseous pollutants taken up by plants include carbon monoxide, nitrogen dioxide, ozone, and sulfur dioxide (Nowak, 1994a). Research has indicated that trees remove significant amounts of pollutants from the atmosphere in urban environments. Trees in Chicago, IL, were found to remove approximately 590 tons of pollution in 1991 (Nowak, 1994a), while in Sacramento, CA, the urban forest removed approximately 330 tons of pollutants from the atmosphere in 1996 (Klaus et al., 1998).

Many gaseous pollutants are taken up into the leaves of trees while particles that stick to the leaf surface are eventually washed off by precipitation and deposited in the soil (Miller, 1997). Trees help reduce the amount of airborne particulate matter by providing a greater surface area to intercept the particles and altering wind patterns and soil erosion to prevent particles from becoming airborne.

18.3.5 Storm Water Mitigation

Trees intercept and assist in the evaporation of rainwater and have been found to reduce storm water runoff by 4%–6% (Sanders, 1986). Other forms of urban vegetation are equally effective at reducing the flow of surface water, thus reducing runoff, erosion, sedimentation, and flooding, allowing storm water to infiltrate into the ground (Miller, 1997). Annual interception of rainfall by the urban forest canopy in Sacramento, CA, was found to be 11.1% of total precipitation in a study using 1992 meteorological data (Xiao et al., 1998). The study indicated that urban forests are effective at reducing runoff during small storms, with coniferous trees being most effective. During larger storms of greater intensity, however, urban forests become increasingly less effective at reducing storm water runoff, reducing their potential as a flood prevention tool (Xiao et al., 1998). Developers are starting to realize that preserving vegetative cover, reducing the amount of impervious surfacing, and providing for storm water filtration and infiltration can save money during development and reduce impacts on the environment (Bodensteiner, 1999).

18.3.6 Provision of Wildlife Habitat

Urban development and fragmentation at the wildland–urban interface can have devastating impacts on wildlife habitat. Habitat areas are segregated, reduced in size, exposed to human activity, and threatened by introduced species (Collinge, 1996). Several aspects of habitat structure must be considered to ensure a suitable home for indigenous wildlife species. These aspects, as indicated by Milligan Raedeke and Raedeke (1995), include patch size, shape, composition, connectivity, and patch dynamics as well as patch viability.

As patch or fragment size increases, so does the amount of adequate habitat. This concept, however, relies on a patch shape that has a small amount of edge interaction with urban development and large amounts of internal space available. Larger habitats with greater structural complexity and diversity have been found to support a larger number of indigenous species (Milligan Raedeke and Raedeke, 1995). As fragmentation due to development continues, it is essential to provide connection between larger fragments to prevent habitat isolation. Natural connection corridors usually exist along creeks, streams, and rivers. Such connections have been described as essential in preventing change in wildlife structure and potential extirpation and even extinction of species (Dunster, 1998).

An important part of habitat complexity in urban landscapes is wildlife trees, defined by Dunster (1998) as trees that provide present or future habitat for the maintenance or enhancement of wildlife. In British Columbia, over 90 species of birds, mammals, amphibians, and insects have been identified as depending to some extent on the existence of wildlife trees. This conservative number does not consider the large number of decay-causing micro-organisms associated with both living and dead trees. Laws to protect wildlife habitat usually require that timber harvesting operations leave a certain percentage of wildlife trees for each ha harvested. Some regulations require resource managers to leave trees in clumps with a minimum of 30 trees per clump and at least one clump for each 8 ha of forest land harvested. Coarse woody debris and dead and decaying logs are to be left to provide additional habitat and a source of nutrients for future forest development. These regulations do not apply, however, when forested land is being converted to urban use. As a result, habitat losses due to development of forested land is not regulated or governed in any way other than through develop agreements with municipal planning agencies.

The challenge for urban forest managers is to provide habitat for wildlife and preserve habitat condition without causing unnecessary hazards or encouraging conflict between urban activities/inhabitants and wildlife. The environmental benefits of wildlife habitat and in particular wildlife trees are often overlooked because these trees and conditions often present a possible safety hazard in the landscape when located in proximity to human activities.

18.4   Water Management

18.4.1   Wetlands, Ravines, Waterways, and Riparian Zones

Naturally occurring wetlands, ravines, waterways, and riparian zones are often protected from urban development. This may be due to the increased cost associated with development of these areas for alternative land-use or because of the recognition of these features as important areas of ecological services. The United Nations Convention on Biodiversity identifies how wetlands play a critical role in maintaining many natural cycles and supporting a wide range of biodiversity. They purify and replenish our water and provide food that feed billions. They serve as a natural sponge against flooding and drought, protect our coastlines, and help fight climate change (United Nations, 2015). Many urban areas around the world have been developed in coastal areas and have resulted in significant damage to wetland and riparian zones. China lost 2883 km^2 of wetlands to urban expansion between 1990 and 2010, of which about 2394 km^2 took place in the eastern regions (Northeast China, North China, Southeast China, and South China). The rate of urbanization-induced wetland loss was 2.8 times higher between 2000 and 2010 (213 km^2 per year) than between 1990 and 2000 (75 km^2 per year) (Mao et al., 2018).

Urban flooding and rising sea levels often associated with a changing climate have become major areas of concerns for both coastal cities and other urban areas. The destruction of natural waterways, riparian zones, and wetlands has resulted in a loss of ecological services associated with flood control. These areas have been covered up or capped with urban infrastructure that is usually impermeable resulting in a significant net increase in runoff during rain events. City planners around the world have recognized this problem and the need to replace hard infrastructure with green infrastructure to provide these critical ecosystem services.

18.4.2　Storm Water Systems

Examples of impervious surfaces in urban areas include rooftops, paved roads, parking lots, and sidewalks. When rainfall strikes these surfaces, it is usually directed to a storm water collection system composed of surface and subsurface channels to carry the excess water to existing natural waterways. Several problems are created by this approach to managing rainfall and excess water. Impervious surfaces do not allow water to infiltrate into the underlying soil and recharge the local water table; they are also often a source of pollutants, commonly from air pollution and automobile use, which excess water suspends as it travels across the surface. The handling systems used to remove the excess water also allow for little water infiltration, increase the velocity of storm water movement, and are costly to install. The polluted storm water is usually directed towards natural waterways through handling systems or existing natural runoff channels. The United States Environmental Protection Agency (EPA) has identified non-point-source pollution as the second most common source of water pollution for lakes and estuaries and the third most common source for rivers nationwide (USEPA, 1994).

As a result countries around the world have implemented storm water mitigation programs to reduce the impact. The United States Environmental Protection Agency implemented widespread *low impact development* (*LID*) standards in the 1990s as a strategy to reduce storm water runoff in urban areas. LID uses a variety of best management practices (BMP's) that include traditional storm water management tools such as retention and detention basins as well as new strategies including bioswales, engineered wetlands, and storm water ponds (USEPA, 2009).

In the late 1990s, Jim Conlin of Scottish Water coined the term *sustainable urban drainage systems* (*SUDS*) in order to bring focus to the new developments in technology regarding storm water management. The first stages of SUDS involved a group of techniques that attempted to mimic natural drainage from an area that would have occurred predevelopment. The intent was to find a more sustainable way to control storm water runoff. At approximately the same time, in Australia, another new term was coined: *water sensitive urban design* (*WSUD*). This parallel approach is a "philosophical approach to urban planning and design that aims to minimize the hydrological impacts of urban development on the surrounding environment". Lloyd et al. (2002) described how storm water management is really a subset of WSUD with the goal of positively affecting flood control, water flow management and improving water quality, at the same time providing methods for harvesting storm water for non-potable uses, to complement more conventional water supplies. Other countries have since developed extensive programs targeted towards reducing storm water runoff. China implemented the *Sponge City Strategy* in 2013/14 in response to problems with urban flooding and rain storm resilience (Fig. 18.4).

All of these strategies represent a significant paradigm shift, with municipal planners now using the concept of ecology to design infrastructure. This has meant not only more environmentally sound systems but also greater efficiency since the new systems have increased adaptability and resilience.

18.5   Agriculture/Urban Agriculture

Agriculture and cities must also co-exist in order to provide a secure food supply to urban inhabitants. Unfortunately, the development of cities has become a threat to agricultural lands in many areas of the world prompting a different strategy for food production. The city of Beijing increased from 4822 km^2 in 1956 to 16,808 km^2 in 2018. This has led to the increased adoption of peri-urban or suburban agriculture. More than 70% of non-staple food in Beijing was produced by the city itself in the 1960s and 1970s and the city continues to be a world leader in Peri-Urban agriculture by advancing the *Agro-Park Concept*. This strategy places agricultural production as an integral part of the urban ecosystem. In addition to the role of urban or peri-urban agriculture to provide food, these developments also provide additional ecological services such as pollutant removal, carbon sequestration, and storm water mitigation. They can also, if managed effectively, help reduce urban air temperatures by absorbing and not reflecting solar energy (Fig. 18.5).

Fig. 18.4   Sponge city park in Fuzhou China (photo credit: T. MacKenzie)

Fig. 18.5   Tea plantation and roadside greenspace in Fujian province, China (photo credit: T. MacKenzie)

Green Roofs/Green Walls

Other spaces within the city represent untapped potential for providing important ecological services. Roof space on buildings represent a significant area of impermeable surface in the urban ecosystem and a primary source of urban storm water runoff. They also represent areas of significant solar reflectance that cause the urban environment to warm up much faster than natural ecosystems. This phenomenon is called the urban heat island effect and was first identified in the late 1960s (Bornstein, 1968).

Despite the negative environmental contribution of urban roof space, these areas represent an opportunity for design and management strategies directed towards reducing their impact and providing ecological services. A green roof is a vegetation system designed specifically to provide vegetative cover on a roof top and the associated ecological services.

There are two common types of green roofs. *Extensive green roofs* are those vegetative communities established on a growing media or substrate that is usually less than 10 cm in thickness. This type of green roof has the benefit of providing vegetative cover with relatively low weight. The establishment of specific plants groups that can adapt to rooftop conditions can reduce roof temperatures, improve air quality, slow down rooftop runoff, and provide habitat in the urban ecosystem. *Intensive green roofs* are systems with growing media or substrate that will support much more extensive root systems and therefore larger and more diverse plant species. These systems provide all the same benefits as extensive roofs and can be used to increase rooftop biodiversity and habitat conditions. This comes at the expense of higher weight that usually requires significant structural engineering in the building below (Fig. 18.6).

Fig. 18.6　Green rooftops in New York (photo credit: T. MacKenzie)

Vegetative systems can also be installed on a vertical plane requiring less urban space to deliver valuable ecological services. *Living walls* can provide growing environments for many adaptable plants and provide tremendous esthetic, ecological, and economic benefits in an urban environment (Shewaka and Mohammed, 2012).

Parks, Gardens, and Green Spaces

The development of cities throughout history can also be correlated with the development of spaces within them for people to gather. As cities became larger and more complex these public spaces evolved as well. With increasing population density living conditions have been shown to decline and citizens demanded improved spaces for gathering, rest, and relaxation. City parks, gardens, and green spaces have been integral parts of cities throughout history providing significant social and ecological services to the people who live there. It has been suggested that that urban parks are essential for liveable and sustainable cities and towns (IFPRA, 2013).

Urban parks have been found to provide significant environmental services similar to those explained for the urban forest with the additions of direct and indirect health benefits, social cohesion, and tourism (Fig. 18.7).

Fig. 18.7   Botanical garden, Fujian Agriculture and Forestry University, Fujian Province, China

(photo credit: T. MacKenzie)

18.6   Planning for Urban Land-Use

18.6.1   Ecosystem-Based and Traditional Land-Use Planning

Traditional land-use planning and regulation can have tremendous impacts on the social characteristics of a community. A comprehensive plan provides guidance for future growth and development by identifying and considering community development alternatives (Elmendorf and Luloff, 1999). The common tools used by municipal planners include comprehensive development plans, zoning regulations, planning ordinances, and development agreements. Planners review and modify land development proposals, facilitate public participation, and assist in collecting, analyzing, and providing information to guide the municipal decision process. Elmendorf and Luloff (1999) note:

In some municipalities, such as Boulder, Colorado and Thousand Oaks, California, land-use planning and regulation coupled with dedicated public involvement, have worked to conserve large green space systems, but these may be the exception rather than the rule. Although land-use planners have long promoted conservation of greenspace, some believe that the benefits of such systems to the quality of life and the environment continue to be ignored.

Improvements to the land-use planning process are needed to reduce the environmental impacts of suburban sprawl. The following changes have been recommended:

The first is to change land use planning to a regional or ecosystem level instead of being based on political borders. Secondly, we must make use of the technological tools that are now available such as satellite imagery, digital demographic information, vegetative and wildlife inventories and geological studies. These can be combined into a single computerized system known as the Geographic Information System (GIS). Finally, the analyses from regional planning ideas and output from any technical assets must be synthesized into public policy decisions. These political decisions will determine the fate of sprawl and at the same time the fate of wildlife. (Nichols, 1999).

Some urban centers have employed basic methods and tools of environmental impact assessment in community-based planning exercises, including scoping, problem characterization, baseline data collection, impact analysis, and the use of matrices, mapping, and GIS (Brugmann, 1996).

18.6.2   Geographical Information Systems (GIS) and Urban Planning

Effective land-use planning requires an accurate and comprehensive assessment of land area. GIS are now used extensively by planning departments to assist in community planning and development. GIS have also been used successfully in the past for processing and updating tree survey data (Widdicombe and Carlisle, 1999; Kane and Ryan, 1998) as well as to assess land coverage and land-use attributes in various cities throughout the world (Pauleit and Duhme, 2000).

As population increases and urban development infringes upon the natural landscape, it is becoming increasingly more difficult to maintain the integrity of natural forest ecosystems at the wildland–urban interface. Scientists, planners, developers, and decision-makers have begun to consider the many valuable roles of vegetation in urban areas and the need for an ecosystem-based approach to managing urban forests. GIS have proven to be an effective tool in supporting an ecosystem-based approach to urban forest management through use in data storage and land-use monitoring (Elmendorf and Luloff, 1999).

18.6.3   Green Infrastructure Planning

GIS have been used successfully to assess urban growth and vegetation structure in large urban areas (Bell et al., 1993). The city of Sacramento, CA, used remotesensing technology and GIS to measure increase in urban growth over a 15-year period and characterize the growth according to vegetative cover patterns. Satellite imagery was used to view the entire urban region and identify the areas of greatest impact. This information could be extremely useful in determining areas at risk from development and areas requiring decisions for future land-use planning.

An ecosystem-based approach to land-use planning and management (Rowantree, 1995) has been found to conserve natural habitats. On a regional scale, GIS have been used successfully to identify large ecologically valuable areas (Elmendorf and Luloff, 1999). GIS play an important role in evaluating natural areas within a community as well. In the early 1980s, the Montreal Urban Community used GIS to evaluate woodlots, streams, and islands on the basis of their ecological value for conservation purposes (Lajeunesse, 1994).

GIS are particularly well suited to comparing spatially defined landscape units by analyzing attribute data. Landscape units with similar attribute data can be merged to create larger landscape units. The merging process is based on the attribute similarity and the spatial configuration of the landscape units making up the area of interest (Colville, 1998).

18.6.4 *Assessing the Benefits of Greenspace*

GIS have allowed urban forest managers to show the benefits of the urban forest to a community. Citygreen[TM] is GIS-based software developed by American Forests (2000) to evaluate the benefits of urban tree cover. Such benefits include carbon sequestration, removal of air pollutants, storm water reduction, energy conservation, avoided emissions, and wildlife habitat (American Forests, 2000). A study in Stevens Point, WI, used Citygreen[TM] software to evaluate the benefits of the urban tree cover. Land-use information was obtained from existing municipal maps to create a base layer in the GIS database. Land-use categories included single-family residential, multiple-family residential, mobile home, commercial, institutional, parks, undeveloped, agriculture, roads, and water. Aerial photographs were examined to develop a land-cover layer to show areas of tree canopy, grass/herbaceous cover, and impervious surfaces. Analysis of the data by Citygreen[TM] provided information regarding open-space distribution, energy savings, and storm water-reduction benefits. This information is considered vital to open-space decision-making and coordinating urban development in a manner that recognizes the benefits of the urban forest (Dwyer et al., 1999).

Conclusion

As urban population continues to increase and cities continue to expand we see a critical need to improve the urban ecosystem by thoughtful assessment, design, installation, and management of green infrastructure. The important ecosystem services provided by green infrastructure must be considered in the context of the contribution of other forms of infrastructure necessary for the function of a city. Tools such as Geographic Information Systems must be used to better determine the impacts of urban development and also the ecological contribution of both natural and manmade landscapes within a city.

Part V

Agroecosystem Management: Issues, Problems and Solutions

Chapter 19
Integrated Pest Management (IPM): From Theory to Application

G. C. Cutler

"Bugs are not going to inherit the earth. They own it now, so we might as well make peace with the landlord."
—*Thomas Eisner*

Abstract Integrated pest management (IPM) is an approach to crop production that is fundamentally grounded in understanding the ecological interactions of insect pests in agroecosystems in order to optimize application of environmentally and economically sound pest management techniques. This chapter overview: the historical development of IPM; major sampling techniques and principles used in monitoring insect pests in agroecosystems: economic principles that drive management decisions: and key aspects of cultural control, insect behavior, plant-resistance, biological control, and chemical insecticide use that should be considered in development of IPM programs.

Learning Objectives

At the end of studying this topic, students should be able to:

1. Define the following terms: pest, integrated pest management, equilibrium position, economic injury level, economic threshold.
2. Explain how insects can injure plants.
3. Describe various techniques of sampling insect pests.
4. Briefly explain the basic economic principles that drive insect pest management decisions.
5. List and explain key aspects of cultural control, insect behaviour, plant-resistance, biological control, and chemical insecticide use that should be considered in development of IPM programs.

19.1 The Importance of Insects

Complex multicellular life on our planet critically hinges on insects. As the most dominant animals of terrestrial ecosystems—the number of species, amount of biomass, and diversity of habitats occupied by other animal taxa pales in comparison—insects fill innumerable ecological niches and provide unrivaled contributions to crucial ecosystem services. These include: pollination of flowering plants; recycling of organic matter; promoting aeration, tilth, and organic composition of soil; symbiosis with other organisms; and general ecological stability through an almost infinite number of food web interactions.

Essential ecosystem services from insects that have driven the functioning of natural ecosystems for millions of years are also fundamental to agroecosystems. Without rich communities of insects to assist us with natural pest control, crop pollination, and recycling of plant and animal waste, most of our crop production systems would soon be non-functional. Indeed, without the help of insects, agriculture as we know it would not even exist.

Despite the many benefits we humans accrue from our co-existence with insects, they clearly can cause harm to us and our possessions. A certain insect in a certain situation would be considered a pest if it *causes annoyance or injury to human beings, human possessions, or human interests.*(Box 19.1)

Box 19.1   How Many Insects Are Pests?

It is important to remember that insects as pests are the exception far more than it is the rule. Of the more than 1,000,000 described species of insects, fewer than 10,000 (<1%) would be considered sporadic or occasional pests, and far fewer still would be considered serious, frequent pests.

Insect pests can impact our lives in a number of ways. As *pests of medical or veterinary importance*, insects can cause discomfort or distress, inflict injury, spread diseases, or cause death. They may do so directly by causing pain and suffering through bites, stings, or venom, and indirectly by being vectors for pathogens or parasites. For humans, this can result in economic losses through medicine and healthcare, as well as losses from stress, absenteeism, and reduced productivity. Likewise, many domesticated animals can be subjected to pain, disease, and discomfort from insects.

Insects can also be pests of our stored products, household commodities, and structures. For example, even in countries like the United States that have good storage facilities and readily available control measures, it is estimated that billions of dollars in post-harvest losses occur due to insects. Insects can cause damage to textiles and natural fibers in our home and can result in billions of dollars in losses to lumber, wood products, and wooden structures.

In this chapter, we are concerned mainly with the occurrence and control of insects as pests for agriculture and horticulture. This is significant because over half of all insects feed on plants, and every plant we know can be attacked by insects. Insect herbivores can injure a plant in many ways. Some insects directly feed on plant tissues with chewing mouthparts, attacking roots, stems, leaves, buds, flowers, fruit, or seeds. Other insect herbivores have piercing-sucking mouthparts, feeding on juices from the phloem and xylem of plants, while in many instances spreading plant pathogens at the same time. Some insects lay their eggs in fruit, causing direct damage as they slice or penetrate into the plant tissue; once infested with the maggot or larvae of an insect, such fruit are either not marketable, or are subject to severe reductions in value or use (e.g., lower value juice vs. fresh market).

It is difficult to accurately estimate crop losses due to insect pest, but several comprehensive reports estimate 10% pre-harvest losses from insects, with further crop losses from insects post-harvest.

19.2   The Development of Integrated Pest Management

Although humans have been trying to control insects since the very beginning of their co-existence—swatting a mosquito is a form of pest control!—it is probably during the emergence of agriculture that insect pests first became a serious threat to human survival. For millennia, we have employed a wide range of tactics to kill or suppress insect populations that attack our crops, ranging from mixtures of natural substance and botanical extracts (e.g., Sumerians 2500 B.C., pre-biblical China, and Egypt), to biological control (1200 B.C. China). The late 1800s and early 1900s ushered in many technological advances, mainly centered on insecticide application equipment, including commercial development of engine-powered sprayers, and airplanes outfitted with pressurized boom sprayers. However, the insecticides applied with these sprayers in the early twentieth century were not very efficient. They were usually expensive, difficult to use, hazardous to apply, and often phytotoxic to the crops themselves. Consequently, growers could not rely solely on insecticides and therefore continued to utilize traditional non-chemical means of pest suppression.

That changed in the 1940s, following the discovery of the insecticidal properties of DDT by Paul Muller in 1939. This discovery ushered in the "insecticide era" (1939–1962). A large number of synthetic compounds that had miraculous activity against insects were developed during this period, largely backed by petroleum companies who expanded into development of agricultural chemicals. These chemical insecticides were so effective at killing insects that most farmers and agricultural entomologists abandoned all other methods of pest control. Insecticides seemed to be a "silver bullet" in the battle against insect pests: they were (are) generally fast acting, reliable, economical, and easy to use. If an insecticide was found to be effective against a pest, it would usually be applied with little regard for potential adverse impacts in the environment, or with little knowledge or concern for pest phenology, density in the field, or damage potential.

Not surprisingly, with this overuse and sole reliance, problems soon became apparent. Insects developed resistance to many products rendering them ineffective; pesticide residues in soil and water increased; and reports of adverse effects on fish and wildlife became commonplace. Applied entomologists realized that reduced and more strategic use of insecticides was needed, and that this had to be done within a framework based on integration of multiple tactics to suppress insect pest populations. It also became clear that all measures to suppress pest populations should be grounded in a comprehensive ecological knowledge of the agroecosystems and the inherent interactions of the players, should pose minimal risk to the environment (soil, air, wildlife, people), and should make economic sense for the farmer. Moreover, rather than trying to "control" or eliminate the pest population—a futile venture—a more tangible and satisfactory goal would be reduction of the pest population to tolerable levels. From this, the concept of Integrated Pest Management (IPM) grew: *A comprehensive approach that uses multiple and combined means and ecological knowledge to reduce the status of pests to tolerable levels while maintaining environmental quality* (Pedigo and Rice, 2009).

IPM is, in many ways, a solution to an ecological problem. The size of a pest population and the damage it inflicts are a reflection of the design and management of a particular agricultural system. Having a better appreciation for the ecology of the agroecosystem—the interplay between the pests and other biotic and abiotic factors in the system—allows us to better understand fundamental questions of why the pest is there, how it arrived there, and why predators or parasites in the complex do not adequately suppress the pest population.

19.3   Sampling and Monitoring for Insect Pests

How does a farmer know if they have an insect pest in their field? The obvious answer is that they must look for it. But there are many ways to look for insect pests, and insect populations likely vary in different parts of a field at different time. Understanding different techniques and principles to find or quantify insect pest populations is a critical component of any IPM program. For some insects, government or commercial agencies may be involved in surveillance or surveys at a national, state, or provincial level, particularly when there is a concern of a newly introduced pest or potential invader; we shall not concern ourselves with pest monitoring at that level for the purposes of this chapter. Rather, we shall briefly describe a few key insect sampling techniques and principles that can be applied by growers or extension specialists at the farm level.

Whatever the sampling technique, it is critical to minimize sampling bias and ensure data quality. This will lead to sound IPM decisions. Key considerations for sampling are:

- *Record keeping*. Whether from week to week or year to year, keeping track of pest populations or plant injury in relation to date, plant development stage, location, weather conditions, and recent field management (e.g., fertilizer or pesticide applications) will prove invaluable in helping understand and predict where and why pest outbreaks are likely to occur.
- *Insect life stage*. As part of record keeping, keeping track of different life stages of pests over a season has ramifications for management decisions. For example, early (smaller) instars do not injure plants as much as late (large) instars, and certain control options can be very effective against early life stages but ineffective against adult insects.
- *Timing of sampling*. In addition to optimizing timing of sampling during the season to coincide with the general phonology of an insect pest, sample timing during a 24-h period can be important. For example, due to factors like temperature or humidity, some pests are inactive at

mid-day, and may be only detectable during dawn, dusk, or at night.

- *Number of sample units.* General recommendations for the number of samples to collect will vary across crops and pest types. When pest populations are very large or very small, relatively few samples units will be needed to decide whether management is warranted.

- *Spatial pattern of sample.* It is important to conduct random sampling for pests throughout an entire field. Such sampling is often conducted in a "W" or zig-zag pattern. Sampling at random locations in a field will minimize bias, e.g. sampling only in convenient locations, such as adjacent to a farm road but not the field interior. At the same time, sampling in known or potential "hot-spots" is recommended. For example, certain pest will be more prevalent in unexposed shaded areas near weed patches.

Common sampling techniques for insect pests are descried below:

In Situ Counts

Sometimes simply counting insect directly where you see them is the most appropriate sampling option. These in situ methods involve no special equipment, although occasionally a hand lens is useful. In situ counts are most effective for large, easily identified insects, and may be limited to certain parts of a plant, e.g. the underside of three randomly selected lower leaves of a plant, or the number of insects per shoot or branch.

Knock-Down

Sometime insects can be effectively removed from their habitat. Though in some cases exposure to chemicals or heat can be useful to remove insect from plants or plant parts, often physical jarring of the plant will suffice. For example, in orchards, a white sheet of specified dimension can be held under a tree branch while the branch is struck a specified number of times. Insects that fall on to the sheet can thereafter be easily counted. Variations of this technique involve shaking a part of a plant into a jar or container to dislodge insects for counting.

Netting

Netting is a widely used and relatively inexpensive technique for sampling insects. Typically, a muslin net is swept through the canopy of a plant(s) and in doing so insects are jarred of the plant and fall into the net. After a prescribed number of sweeps, the contents of the net can be dumped into a container, and the insects counted.

Trapping

Many trapping techniques have been developed for insect pests. Irrespective of the trap used, for a traps to be effective ①the insect must move, and ②the trap must hold captured insects for a prescribed amount of time. While some traps are passive and have no visual, chemical, or auditory cue that attracts the insect, most traps are designed to be attractive to an insect pest in one or more ways.

An example of a passive trap for collecting insects is a pitfall trap—a simple cup or container that is pitted into the soil that can capture insects that actively walk on the soil surface. These types of traps are especially good at collecting beetles, many of which may be beneficial predators of insect pests in agroecosystems. To capture flying insects, Malaise traps may be effective. These are mesh, open-fronted tents that intercept flying insects. The Malaise trap is designed such that insects that fly into the tent are collected in container after the move upward to the top of the tent.

Traps are often designed to be visually attractive to insects. For example, most aphids are attracted to the color yellow, and can be trapped on a yellow card coated with an adhesive (e.g., Tanglefoot®) when placed in the plant canopy. Likewise, red spheres coated with adhesive hung in an orchard can capture fruit flies (Tephritidae) that are attractive to ripe, red fruit.

Other traps incorporate smells that are attractive to insect pests. Volatile chemicals that mediate insect behavior are called semiochemicals. These will be discussed in more in the Insect Management Options section of the chapter.

19.4　Economics of Insect Pest Management

Let us say you are a farmer and you find an insect pest in your field. We understand that more individual pests generally mean more potential injury to plants, and subsequent economic loss. But how do you know whether you should take action to manage the pest? As a farmer trying to make a living and earn an income, decisions regarding when and how to manage an insect pest population should, first of all, have a basis in economics. That is, what level of pest population or pest damage will cause economic loss to the farm operation? Without an appreciation of the economic costs and economic benefits of a particular management decision, a farmer risks spending more on management to reduce an insect population than the value of the crop he/she is trying to protect.

While a small pest population will usually result in low economic losses, at some point a pest population will reach a size where the cost to manage the population equals the amount of economic loss (actual or potential) the pest population causes. This "break even" point is called the *economic injury limit* (*EIL*). This means below the EIL, it is not cost-effective to control a pest population because treatment costs exceed the value of crop losses caused by the pest, whereas above the EIL treatment costs are compensated by an equal or greater reduction in real or potential economic losses caused by the pest. This concept is characterized as the *gain threshold*, which is a quotient of management costs/market value. For example, if corn is valued at $4 per bushel, and an insecticide costs $20/ha to apply, the gain threshold is:

$$\text{Gain threshold} = \frac{\$20 \text{ per ha}}{\$4 \text{ per bushel}} = 5 \text{ bushels per ha}$$

This means at least 5 bushels of corn per ha would need to be saved with the insecticide application in order to justify the costs of applying that insecticide. Related to this, the EIL is a function of the gain threshold and a measure of loss per insect:

$$\text{EIL} = \text{gain threshold/loss per insect}$$

A concept related to the EIL is the *economic threshold* (*ET*), or the action threshold—a point below the EIL where a decision is made to implement an action (e.g., insecticide spray) to reduce the pest population. Both the EIL and ET are usually expressed in units of insect density (e.g., 10 insects per 20 sweeps; 2 insects per trap per ha; 5 insects per 5 m row of plants), but may also be expressed as a measurement of injury (e.g., 5% defoliation per plant). The ET is often set conservatively, for example, at 75% the EIL.

19.5   Management Strategies

Many pest management strategies and tactics have been used in IPM programs. Described below are some of the major practical approaches that can be used by farmers for pest suppression in agroecosystems.

19.5.1   *Cultural Control*

Pest management tactics that change the agroecosystem environment or habitat to reduce pest populations or damage are known as "cultural control" techniques, so named because they involve modification of standard management practices. Because these methods change the environment of the pest, they are also characterized as ecological management. For these methods to be effective, we need to understand a pest's ecological requisites: the what, where, and when of the pest's habitat, shelter, food sources, reproduction, natural enemies, or physical environment requirements. With this understanding, a farmer can exploit "weak links" in the insect's life cycle or behavior. Brief descriptions of some effective cultural control tactics follow. The benefits of these tactics need to be balanced against potential adverse impacts on the agroecosystem, e.g. removal or habitat for natural enemies or pollinators, soil erosion.

19.5.1.1   Destruction or Removal of Alternate Hosts or Habitats

Many pest species breed or overwinter in debris or alternate hosts in or around a crop field or crop storage facility. If these insect habitats are "cleaned up", a farmer can effectively suppress or remove a significant source of the pest population. Examples of good sanitation practice are: removing animal waste from barns to oviposition sites for flies; removing fruit (which may contain insects or still serve as suitable reproduction sites) that have dropped in an orchard and feeding to animals; removal of weeds around a field that can harbor pests; plowing under crop residues that remain in a field; or cleaning up spilled grain, fruit, or vegetable around a storage facility.

19.5.1.2 Irrigation and Water Management

Too much or too little water can be detrimental to a pest. For example, anaerobic conditions created through temporary flooding of fields have been shown to reduce wireworm populations, and insect pests of cranberries. Periodic irrigation with overhead sprinklers can disrupt moth mating, oviposition, or larval development in vegetable crops, and can reduce infestations of two-spotted spider mite in tree fruit.

19.5.1.3 Tillage

A large portion of insects spend part of their life cycle in or on soil. Tillage of soil can modify soil texture, moisture, or temperature to the detriment of insect development, and expose previously subterranean insects to natural enemies like birds.

19.5.1.4 Crop Rotation

Monoculture methods (planting a single crop several years in a row in the same field) allows a pest population grow to damaging levels. Although it is often used as a means of crop nutrient management, crop rotation creates a discontinuous food supply for a pest and can be a very effective means of breaking or hindering its life cycle. For example, a rotation of corn with soybean, oats, or wheat over a 2–3 year cycle is a fundamental management tactic for corn rootworm (*Diabrotica* spp.) in temperate zones. Crop rotation works best against pests that: attack annual crops; have a narrow host range; and have limited mobility.

19.5.1.5 Disrupt Crop-Insect Synchrony

The life cycle of a pest is often synchronized with that of a plant. In some cases, it is possible to grow crops "out of phase" with that of the pest by altering planting dates, thereby reducing susceptibility of the plant to injurious pest life stages. This tactic is effectively used in Hessian fly (*Mayetiola destructor*) management. By tracking temperature during the season as a means to predict fly development, planting of winter wheat can be delayed until the threat of female flies (who lay their eggs in developing wheat) has passed. For other crops, it may be possible to plant earlydeveloping varieties that can be harvested before injurious pest stages become prevalent.

19.5.1.6   Push–Pull Cropping

It is possible to plant crops spatially so that they affect the dispersal behavior of insect pests, "pushing" them away from the crop of value or "pulling" the pest into a crop/planting of less or no economic value. "Trap cropping" entails planting small areas of attractive alternate crops (the "trap") near the crop to be protected, leading to a concentration of the pest off the main crop. For example, strips of alfalfa interplanted in cotton are attractive to *Lygus* bugs, thereby concentrating ("pulling") the pest in the alfalfa, which can be managed before the bugs move into the cotton (Pedigo and Rice, 2009). At the same time, growing different crops in a single location minimizes monocultures and may lower the attractiveness of the whole area, "pushing" the insects to an alternate habitat.

19.5.2   Behavioral Control

The behavior of many insects is mediated by volatile semiochemicals or infochemicals that serve as a "language" to mediate interactions between organisms. These semiochemicals may attract or repel insects, stimulate or inhibit feeding, or provoke flight or inhibit it. We can, therefore, use our knowledge of insect semiochemicals to manipulate insect behavior in monitoring or managing insect pest populations.

Pheromones are semiochemicals that facilitate communication between the members of a single species, and include sex attractants, trail marking compounds, alarm, and aggregation substances. Pheromones are probably the most used semiochemicals in insect pest management. After synthesis in a laboratory, pheromones for a specific behavior of a specific species can be impregnated into slow-release dispensers (rubber, hollow fibers, or rope) and placed in traps of various designs. At low densities these pheromone traps can be valuable monitoring tools, whereas at high densities they can be effective for mass trapping of sexually active adults. When placed in a field at relatively high densities, sex pheromones can be useful in mating disruption. For example, with a high concentration of sex pheromone in the field, everything will smell like a receptive female to a male moth. This will inhibit the ability of the male to target a female moth, such that the male moths eventually wear themselves out or become habituated (non-responsive) to the odor. By interfering with mating, this technique can suppress the pest population.

19.5.3   Plant Resistance

Plants have evolved for millions of years with plant-feeding insects and developed a wide range of physiological and morphological characteristics that serve as a basis for defense to insect herbivory. In nature, selection of these defensive traits in plants occurs through natural selection. However, human intervention through plant breeding or biotechnology has allowed us to produce plants with genetic traits (or combinations of traits) that reduce susceptibility to attack or injury. In general, there are three main modes through which plants may have reduced susceptibility to insect pests:

1. *Non-Preference*. Also known as "antixenosis", non-preference involves exploitation of plant characteristics that detract the insect from the plant. The non-preference for the plant can be chemical in nature, such that the insect is deterred from feeding, or it can be morphological, in the form of hairs, waxes, or a thick, tough epidermis that provides an undesirable feeding, oviposition, or habitable substrate.

2. *Antibiosis*. This involves impairment of insect metabolism, usually after a pest feeds on the crop. Antibiosis frequently involves chemicals produced by the plant that induce various symptoms in the insect, including mortality, reduced rates of growth, reduced size, reduced fecundity, shortened life span, altered behavior, or morphological malformations.

3. *Tolerance*. Some plant genotypes are able to "tolerate" injury better than others. These tolerant cultivars suffer less damage than susceptible cultivars when attacked by pests, due to general overall vigor, wound healing, new growth, or resource partitioning. For example, corn that has thicker, stronger cornstalks will suffer less wind breakage after attack by European corn borer (*Ostrinia nubilalis*) larvae that burrow into stalks. Other corn genotypes have enhanced ability to replace roots damaged by western corn rootworm (*D. virgifera*).

Crop cultivars with pest resistance can be developed through breeding either using conventional methods or transgenic methods. In conventional plant breeding, typically a desirable trait (e.g., production of large amounts of allelochemicals that deter pest feeding) is identified in the resistant/tolerant cultivar, and this trait is transferred to another cultivar by crossing (breeding). Progeny of the cross are evaluated for the presence of the resistant trait. With repeated crosses over several generations, the plant breeder can ensure consistent and high levels of expression of the desired trait, and minimal expression of undesired traits.

Though effective, traditional breed methods can be a slow (typically 10 years or more) and there is much uncertainty around ensuring consistent, high expression of the trait(s) of interest. Biotechnology allows us to insert genes from one organism into the cells of a completely different species. In insect pest management, this has been most successful in development of transgenic "*Bt* crops", including cotton, potato, and corn. *Bacillus thuringiensis* (*Bt*) is a common soil bacterium with several genes that produce delta-endotoxin proteins that are toxic to insects. There can be a high level of delta-endotoxin specificity: some are highly toxic to beetles (and far less toxic to other groups of insects), others are selectively toxic to caterpillars, and others still are selectively toxic to flies. *Bt* toxins can be cultured from bacteria in the laboratory and subsequently formulated for sprays in the field but such sprays have limited persistence due to photo-degradation from the sun's UV light, and wash off in rain. To address these problems, several crops have been genetically engineered to incorporate *Bt* delta-endotoxin genes into the crop genome. Specific *Bt* genes are inserted into the plant embryo (usually accomplished with *Agrobacterium*, a bacterium that acts as a "natural vector" for horizontal gene transfer in plants), resulting in a mature plant that expresses the delta-endotoxin gene in all its cells. When herbivorous pests feed on the plant, they ingest the toxins proteins, which results in the eventual death of the insects. Being incorporated into the plant's genome, deltaendotoxin proteins are protected from photo-degradation, and are produced continually over a season, extending the duration of protection from pests.

19.5.4   Biological Control

Biological control involves strategies and tactics that manipulate *natural enemies* to kill, decrease the reproductive potential of, or otherwise reduce the numbers of pests. When used in pest management, natural enemies are often referred to as *biological control agents* or *biocontrols*. Biological control is one of the oldest and most important methods of insect pest management. Indeed, all pests are affected by biocontrols that often go unnoticed as they maintain pests at a population equilibrium position that is of sub-economic importance. If acclimated to the target area and left relatively undisturbed (e.g., not eliminated by indiscriminant insecticide applications), natural enemy populations can become permanent pest controls in an agroecosystem. There are three principle means by which biological control operates:

1. *Conservation biological control* aims to optimize survival and/or effectiveness of natural enemies that already exist in the agroecosystem. This can be achieved through two principle means:

   (a) Reducing insecticide applications or using "selective" products to minimize or avoid adverse impact on natural enemy populations.
   (b) Provide natural enemies with habitat. For example, farmers can ensure they have adequate hedgerows, forests, floral resources, or host plants around their fields. Depending on the agricultural system, strategies such as less mowing in orchards, plant cover crops, reducing tillage, using less herbicide, or providing mulches may be very effective for maintaining habitats that are important for natural enemies.

2. *Augmentative biological control.* Sometimes natural enemies naturally occur in an agroeco-system but are unable to survive or persist at levels sufficient to result in significant suppression of a pest population. It may, however, be possible to rear these biocontrol agents in large numbers and periodically release them into an agroecosystem. Augmentative biocontrol may be achieved by either release of relatively small numbers of natural enemies in critical locations where progeny of the natural enemy are also expected to have an impact (= *inoculative releases*), or release of very large numbers of natural enemies with immediate impacts and no expected significant effect of progeny of the natural enemy on the system (=*inundative release*).

3. *Importation biological control.* In any given agroecosystem, many pest species are exotic, i.e.

they originated from another region or country. In such cases, it may be effective to go to the country of origin of the pest and seek out effective natural enemies that have co-evolved with them. Following extensive pre-release evaluation to insure that the ecology and host range of the potential biocontrols agent are compatible with the new community, the natural enemy is reared and released in the new habitat in hopes it will become established and suppress the pest population. Because one of the earliest documented biological control successes relied on this importation strategy, the introduction of vedalia beetle (*Rodolia cardinalis*) to California from Australia to control cottony cushion scale (*Icerya cardinalis*), which was accidently introduced to California, importation biocontrol is often referred to as "classical biological control".

Numerous types of biological control agents are used in IPM. Brief descriptions of the most important biocontrol agents follow.

- A variety of *predators* can prey on insects, including birds, mammals, fish, and other arthropods like spiders. But for practical biocontrol of insect pests, predaceous insects and mites have been most important. Beetles (Coleoptera), bugs (Hemiptera), lacewings (Neuroptera), and some flies (Diptera) and ants (Hymenoptera) are usually most important. Insect predators are effective because of their rapid kill, they consume many individual pests, immature and adult life stage may be predaceous, and synchronizing the life cycle of the predator with the prey (pest) is often not a problem.
- A *parasitoid* is an organism that lives and develops at the expense of a host, and eventually kills the host. Many flies and wasps (Hymenoptera) are important parasitoids for biological control. Adult parasitoids are free-living and mobile, while immature life stages develop on or within a single insect host. Only female parasitoids directly contribute to biological control as only females search for hosts. Parasitoid are frequently species/family and stage specific, e.g. certain wasps they will only attack eggs of certain species of caterpillars.
- Some *bacteria, fungi*, and *viruses* are entomopathogenic, meaning they are pathogenic to insects. Some of these may be formulated such that they can be mixed with water and sprayed on to crops for effective in biological control of certain insect pests. Good examples of pathogens are:

  - *Bacillus thuringiensis* (*Bt*). This is probably the most widely used biocontrol agent. As described above, this gram-positive bacterium produces a toxic proteinaceous crystal (delta-endotoxin) that is activated when ingested by insects. *Bt* can be mass produced in large fermentation vats and formulated as dusts, liquid, or granules, providing flexibility for a variety of applications.
  - *Beauveria bassiana* and *Metarhizium anisopliae* are examples of entomopathogenic fungi that have been formulated for applications in pest management. Fungal spores that make direct contact with insect cuticle will germinate and colonize the insect's hemocoel, sometimes producing toxins.
  - *Nuclear polyhedrosis viruses* and *granulosis viruses* are the best known groups of viruses

that infect insects, causing the integument and internal tissues to "melt away" into a liquefied blob. Successful applications of viruses as biological controls have been against forest insect pests (e.g., pine sawflies, tussock moths, gypsy moths) and some orchard pests (e.g., codling moth).

— *Entomopathogenic nematodes* from the genera *Steinernema* and *Heterorhabditis* carry pathogenic bacteria that are released after the nematode enters the insect, killing the insect in 24–48 h. These nematodes can be mass cultured in laboratories and formulated to facilitate easy mixing with water and application with regular spray equipment. They are particularly effective against soil-dwelling insect pests.

19.5.5 Insecticides

The chemical insecticide era came into being with the discovery and widespread use of DDT (Dichlorodiphenyltrichloroethane) in the mid-1940s. Many chemical insecticides have since been developed and they are undoubtedly indispensable in maintaining current levels of health, nutrition, and quality surroundings. Insecticides are a regular component of most pest management programs and have multiple features that farmers appreciate. Chemical insecticides are:

- Fast acting, often killing the pest within minutes or a few hours of exposure.
- Generally reliable, giving consistent pest suppression (see below for discussion on insecticide resistance).
- Often broad spectrum in that a single active ingredient can be effective against multiple pest species.
- Versatile in that they can be formulated and applied in a variety of ways to suit different needs.
- Relatively economical and easy to use, easily being mixed with water and applied with hand-held or tractor-mounted sprayer equipment.

Despite these benefits, overuse, misuse, and abuse of insecticides have led to widespread criticism of chemical control and, in some cases, have resulted in adverse effects on human health and the environment. Though use of insecticides does come with inherent hazards and risks (as do all management options), they can be used safely and can be effectively incorporated into a multi-faceted IPM program.

19.5.5.1 Insecticide Exposure Routes

Insects can be exposed to insecticide in a number of ways, and insecticides may vary in their efficacy depending on the exposure route. Insecticides that have *direct contact* activity are absorbed directly through an insect's body, whereas other insecticides are *stomach poisons* and have to be ingested; many insecticides have both direct contact and ingestion activity. *Systemic insecticides* operate as stomach poisons, but when applied to seeds or soil can be incorporated into the vascular tissues of plants such that only herbivores of the plant tissues are targeted. When targeting pests in enclosed spaces (e.g., greenhouse, warehouse, grain bin, home) or soil, fumigant insecticides that have high volatility and penetrability through the insect's respiratory system can be highly effective.

19.5.5.2　Insecticide Formulations and Types

Pure insecticide is never applied to a crop. The *active ingredient* (AI), which is the compound that elicits the toxic effect on insects, is always formulated with other components. Technical grade insecticide (usually 90%–99% AI) is usually too toxic, unstable, or volatile to be applied to a crop on its own, so it is always mixed with *adjuvants*to improve the performance, safety, or handling characteristics of the insecticide product. The adjuvant "recipe" is proprietary information and these components are usually listed on the insecticide as "inert" ingredients. These inert ingredients may constitute 90%–95% of commercial insecticide products. The mixture of technical grade insecticide with adjuvants gives the insecticide *formulation*. Depending on the target pest and commodity to be protected, insecticide can be formulated any number of ways, including granules or pellets, dusts, wettable powders, soluble powers, emulsifiable concentrates, or aerosols.

Many kinds of active ingredients have been developed. The most effective and practical ones have an organic skeleton (i.e., they contain carbon atoms with covalent bonds) and have been synthetically produced in the laboratory. Some insecticides have active ingredients that are *botanical* in origin in that they are based on extracts from plants; plants have, over millions of years, evolved through natural selection to produce chemicals that are toxic to herbivores that try to eat them, and these chemicals can be isolated from plants and formulated into practical insecticides for agriculture. Indeed, several synthetic insecticides are essentially analogs of naturally occurring insecticidal compounds.

Insecticides are generally classified based on their mode of action, with various chemical sub-groups and active ingredients derived from this. Insecticides may act as nerve poisons (e.g., interference with neuro-transmitters, ion channels of neurons, or muscle action), growth regulators (i.e., interfere with development), inhibitors of biochemical processes (e.g., lipid biosynthesis or energy metabolism inhibitors).

19.5.5.3　Insecticide Resistance

In addition to concerns of insecticides on non-target organisms (natural enemies, fish, birds, wildlife, farmers, consumers), preventing development of *insecticide resistance* is a major challenge. Resistance development is a product of evolution and natural selection. That is, insecticide-resistant populations evolve because some individuals in the pest population were able to survive exposure to the insecticide (i.e., survival of the fittest!), and subsequently reproduced, thereby passing their beneficial resistance traits to their offspring. Usually the traits that render a pest population resistant to an insecticide are biochemical in nature. For example, individuals that tolerate the insecticide may have greater amounts of certain enzymes (e.g., esterases, hydrolases, transferases, and oxidases) that denature the insecticide active ingredient before it can elicit toxic effects. Other times, traits that reduce insecticide susceptibility may be physical (e.g., thicker cuticle that slows penetration of the insecticide into the body of the individual) or behavioral (e.g., some mosquito populations have evolved to learn to avoid insecticide-treated walls).

There are several ways a farmer can avoid or manage insecticide resistance. They all rely on reducing the selection pressure (i.e., exposure to the insecticide) for development of resistant populations. Some key resistance management strategies are:

- *Using less insecticide.* This is the only resistance management strategy that absolutely works, since it is the only strategy that guarantees reduced selection pressure.
- *Increase non-chemical mortality.* The greater the use of techniques that do not involve insecticides—e.g. cultural control and biological control—to suppress the pest population, the less insecticide has to be used, thereby reducing selection for resistant populations.
- *Alternate different chemicals.* Repeated use of insecticides that have the same mode of action increases selection for pest populations resistant to that mode of action. Alternating applications of insecticides with different modes of action across time and/or space alleviates selection pressure for any one kind or group of compounds.
- *Avoiding persistent compounds.* Persistent insecticides will unnecessarily continue to select for resistance even when populations are below economically important levels.
- *Providing refugia.* "Insecticide-free" zones are critical for retaining individuals with insecticide-susceptible genotypes in the population. When these individuals reproduce with other susceptible individuals or resistant individuals, the overall proportion of insecticide-resistant phenotypes in the population is decreased.

19.6   Insect IPM Case Study

Colorado Potato Beetle (*Leptinotarsa decemlineata*)

Colorado potato beetle (CPB) is one of the most important insect pests of potato in North America and in many potato-growing areas of Europe, China, and other parts of Asia. The combination of a flexible life history and a remarkable ability to develop resistance to insecticides makes this insect a challenge to control. CPB is considered one of the top ten agricultural insect pests in the world. It has been suggested that CPB has had a more profound effect on the development of crop protection methods and application equipment than any other insect pest of agricultural crops.

CPB is an excellent pest to use as a case study for insect IPM in that good management relies on a thorough understanding of the insect and the cropping system, which can optimize use of the numerous technologies and techniques that have been developed to manage this insect. The table below illustrates how biological and ecological knowledge lends itself to good tactics and tools for this pest. Bear in mind that no single tactic or tool is expected to eliminate a CPB outbreak, but an integrated approach can effectively reduce pest populations below economically important levels (Table 19.1).

Table 19.1　How biological and ecological knowledge of Colorado potato beetle (CPB) provides insight into tactics and tools for its management

Biological and ecological knowledge	Tactics and tools for management
CPB adults overwinter in soil, often hedgerows around fields, and invade new potato fields in spring	Beetle overwintering diapause success can be reduced by removal of insulating snow layers. Milner et al. (1992) applied straw mulch to overwintering sites and then removed it along with the layering snow in January. This greatly reduced soil temperature, resulting in 20% reduced beetle survival
CPB adults often walk into potato fields in spring	Digging trenches along a field edge and lining them with plastic can intercept migrating CPB and reduce populations immigrating into fields by 50% (Boiteau et al., 1994)
Invasion usually limited to 1 km radius	Because CPB adults often walk into fields and usually do not move more than 1 km from their overwintering site, crop rotation is an effective and relatively easily implemented cultural control technique. Separation needs to be 0.3–0.9 km between rotated fields, which can reduce adult densities 96% (Wright, 1984) and egg mass densities by 90% (Lashomb and Ng, 1984)
Females lay eggs over 4–6 week period, third and fourth instar larvae are most damaging	Monitoring for adult presence/damage, egg laying and hatch, and larval development is key for effective management; the conspicuous nature of the different life stages makes this a relatively easy task. This will, for example, enable targeting of less damaging life stages (first and second instar larvae) with selective products like *Bacillus thuringiensis* (*Bt*) sprays or insect growth regulators (e.g., chitin synthesis inhibitors), thereby avoiding development of more damaging third and fourth instar larvae
Natural enemy population dynamics	A large number of natural enemies of CPB have been identified, including many species of predatory beetles and bugs, parasitoid flies and wasps, pathogenic fungi and bacteria, and parasitic nematodes. Although augmentative biological control is usually not practical, many of these natural enemies will exist in or around fields. Efforts to understand the dynamics of CPB natural enemy populations, and to conserve or boost populations around fields (e.g., habitat management, using selective insecticides), will likely be will rewarded with reduced pest pressure
Resistance to insecticides is widespread	Heavy dependence on insecticides, coupled with an extraordinary biological ability of this insect to evolve resistance, has resulted in a history of CPB management marked by repeated incidents of rapid development of resistance and the exhaustion of nearly all insecticide options in many important potato production regions. Best practices such as regular monitoring for resistance (e.g., insecticide "dip tests" with emerged adults), and employing multiple resistance management techniques will help ensure the viability of chemical insecticides as an option in an IPM program for CPB

Summary

Integrated pest management (IPM) is rooted in a total agroecosystem approach that relies on leveraging our understanding of agroecology to optimize the use of a variety of tactics to suppress pest populations to sub-economic levels. IPM provides farmers a practical framework to generate high yields while maximizing their profits and maintaining environmental quality. The concept of IPM is not a new one. For generations, farmers have used all the tools in their pest management toolbox but the advent of chemical insecticides often resulted in the abandonment of other approaches. Only when the limitations and risks of excessive pesticide use became apparent did we find ways to adapt "old ways", and increase efforts toward discovery, innovation, and implementation of more biorational pest management alternatives. Minimizing inputs and maximizing proactive biointensive strategies and tactics to manage pests allow growers to produce premium, profitable products in a sustainable way.

Chapter 20
Organic Agriculture: A Model for Sustainability

A. M. Hammermeister

"Organic agriculture is a model of food production guided by principles and standards of sustainability and supported by growing consumer demand."
—Andrew M. Hammermeister

Abstract Organic agriculture is one of the fastest growing sectors of agriculture. It is rooted in the belief that healthy soil is the foundation of healthy plants, livestock, and people. Organic agriculture has become an internationally recognized system of production that is based on the principles of ecology, health, fairness, and care. It is guided by production standards that are regulated in many countries of the world, and its rapid growth is driven by consumer demand. Organic production is truly global in nature, including small- and large-scale farmers, intensive and extensive organic production practices, technologically advanced and constrained production, and a wide range of growing environments from tropical to arid to sub-arctic. With its guiding principles and standards, organic agriculture offers a unique model to guide us in our movement towards sustainability of food systems.

Learning Objectives

After reading this chapter, a student should be able to:

1. Define organic agriculture.
2. Identify and explain the principles of organic agriculture.
3. Distinguish between organic regulations, general management standards, and the permitted substances list.
4. Identify general practices encouraged by organic agriculture and those that are restricted.
5. Discuss the significance of organic agriculture as a global movement.
6. List and describe the challenges of organic agriculture.
7. Explain what is meant by process- or outcome-based production system and describe how organic agriculture fits within those definitions.

20.1   Overview of Organic Agriculture

As seen from previous chapters in this text, humanity is living not only in an incredibly interesting but also a challenging time. Agriculture has enabled the development of civilizations with social structure, education, science, and technological development. The rapid growth and spread of human populations around the world have required tremendous advances in agricultural production, food processing, and food distribution. While these advances have tremendously improved the quality of life of many in the world, differences in politics, climate, and resources have resulted in an uneven distribution of wealth and technology on both local and global scales. In the past, traditional farming systems were more diversified, growing a mix of crops and livestock at a smaller scale, often for local or regional markets. The development and industrialization of processes for synthesizing and transporting fertilizers and pesticides through The Green Revolution has resulted in their widespread adoption by farmers around the world. However, economies of scale enabled by mechanization and automation have led to many farms specializing in as few as one crop on a larger land base or concentrated systems of single-species livestock.

While specialization and expansion of agriculture has certainly enhanced global food sufficiency, it has also led to many other challenges. Agriculture is regarded as a primary contributor to biodiversity loss, contamination of freshwater through nutrient and pesticide runoff or leaching, soil degradation, antibiotic resistance, and issues of animal welfare. Educated consumers are showing increasing interest and concern over the impacts of modern food systems. But is there an alternative to intensive agricultural production? Consumers, producers, and policy makers are seeking sustainable and healthy alternatives in food production. Organic agriculture has become increasingly recognized as an alternate system for sustainable production. This chapter will provide a brief overview of organic agriculture and briefly discuss opportunities, challenges, and constraints around it.

20.2   Definition, History, and Principles of Organic Agriculture

IFOAM Organics International (www.ifoam.bio/en) is the international body that brings together global organic movement to ensure there is consistency in the values, principles, and standards relating to organic agriculture. IFOAM Organics International defines organic agriculture as:

> Organic Agriculture is a production system that sustains the health of **soils, ecosystems** and **people. It relies on ecological processes, biodiversity and cycles adapted to local conditions**, rather than the use of inputs with adverse effects. Organic Agriculture combines **tradition, innovation** and **science** to benefit the shared environment and promote **fair relationships** and a good **quality of life** for all involved. (IFOAM Organics International)

This definition is rooted in the four principles of organic agriculture briefly described by IFOAM Organics International as follows (IFOAM Organics International, Fig. 20.1):

The "Precautionary Principle" is inherently part of the organic philosophy even though it is not shown among the core principles of organic agriculture. While the concept of the precautionary principle has evolved over time, it can be defined as:

> When human activities may lead to morally unacceptable harm that is scientifically plausible but uncertain, actions shall be taken to avoid or diminish that harm. (COMEST, 2005, p. 14)

Within this definition it is implied that caution should be used to avoid risk when adopting practices in agricultural even if the risk has not yet been demonstrated scientifically. From an organic perspective, for example, the use of genetic engineering may produce an unforeseen risk to the environment or human health that has not been documented scientifically, and thus alternate practices should be used for pest control instead of using synthetic pesticides.

The Principle of Health.   The Principle of Ecology.   The Principle of Fairness.   The Principle of Care.

Fig. 20.1 The four principles of organic agriculture

Principle of Ecology—Organic Agriculture should be based on living ecological systems and cycles, work with them, emulate them, and help sustain them

Principle of Health—Organic Agriculture should sustain and enhance the health of soil, plant, animal, human, and planet as one and indivisible

Principle of Fairness—Organic Agriculture should build on relationships that ensure fairness with regard to the common environment and life opportunities

Principle of Care—Organic Agriculture should be managed in a precautionary and responsible manner to protect the health and well-being of current and future generations and the environment

History

The concept of organic agriculture emerged as a result of issues in soil quality, food quality, and social life arising from the adoption of chemical agriculture (Vogt, 2007) While others had investigated the ecology of agroecosystems, Sir Albert Howard is generally credited with being the founder of modern organic agriculture through his book "An Agricultural Testament" (Howard, 1943) and inspiring other pioneers of the movement such as Lady Eve Balfour in the U.K. and J.I. Rodale in the USA (Vogt, 2007). These pioneers of organic agriculture essentially recognized the interconnectedness of soils, plants, and human/livestock health and found that degradation of the soil would result not only in declining productivity but also in declining nutritional value of crops for livestock and people. Thus, healthy soils lead to healthy crops and healthy livestock and people. Sir Albert Howard recognized that the maintenance of humus in the soil was critical for maintaining soil quality and that this in turn could be enhanced through additions of compost derived from plants and manure. Coupled with this was an increasing awareness of soil biology and recognition that it had an important role in influencing soil health. Sir Albert Howard emphasized a holistic approach to agriculture rather than a reductionist approach. Lady Eve Balfour, one of the first women to study agriculture in the UK, established what is regarded as the first long-term study comparing organic and conventional farming systems. In the USA, widespread soil erosion drew concerns of soils scientists and ecologists who were concerned about soil protection. These concerns coupled with a "back-to-the-land" movement led to the promotion of organic farming and gardening through the likes of Jerome I. Rodale who started the magazine "Organic Gardening and Farming" in 1942. Intensive use of chemicals became the norm in the USA until Rachel Carson's book "Silent Spring" (Carson, 1962) stimulated the beginning of the environmental movement and concerns over pesticide use (Heckman, 2005).

Through the 1970s to 1990s, growing consumer consciousness of issues in conventional agriculture led to growing consumer demand for organically grown products and a willingness to pay a price premium. The International Federation of Organic Agriculture Movements (IFOAM) established in 1972 with a recognition that there needed to be global coordination of organic agriculture and development of a common standard to enable not only consistency to the consumer, but also scientific analysis of organic systems and comparisons across countries (IFOAM Organics International). As consumer demand continued to grow for organic product, governments at a provincial/state and then national levels began to regulate organic agriculture in order to maintain public trust and allow cross-border trade. While standards vary to some extent among countries, the core production practices and permitted inputs are largely the same.

While this history of organic agriculture has emphasized its development as a regulated system and marketable brand, it is also important to recognize that many forms of traditional agriculture have existed for centuries, if not millennia, without the use of modern technologies and inputs. Sir Albert Howard studied these traditional practices in India and used this as a foundation for his work in "*An Agricultural Testament*".Today, India is the home of the largest number of organic farmers in the world as their traditional practices were easily transitioned to organic agriculture (Schlatter et al., 2020; Fig. 20.2).

Fig. 20.2   India is home to the largest number of organic producers in the world

(photo credit: A. M. Hammermeister)

This small-scale farm is located in a terraced landscape in the lower Himalayas of India

20.3　Organic Standards and Regulations

As mentioned above, the rapid growth of consumer demand for organic products led to international trade and thus a niche market opportunity. Many countries began by developing a national standard and then a system of inspection to ensure that farmers complied with the standard. However, it became clear that the integrity of the "organic product claim" of producers, processors, traders, and retailers needed to be protected by a regulated system that ensured compliance with organic standards. While "standards" describe the practices and inputs that may be required, permitted, or prohibited in organic operations, government regulations create a framework for enforcement. As such, over half of the countries in the world that report production of organic products have developed or are in the process of developing legislation to regulate organic agriculture (Huber et al., 2018).

20.3.1　Organic Regulations and the Certification System

The system of regulation can vary by country but will have similar components; here Canada will be used as an example. In Canada, organic agriculture and food products (including feed, but not textiles, pharmaceuticals, or cosmetics) fall under Part 13 of the Safe Food for Canadians Regulations (Government of Canada, 2019). Non-food products may become certified as organic but are not regulated under this act. The regulations require multiple layers of oversight. An organic operator (ex. producer or processor) must apply to an accredited certifying body for organic certification every year. The certifying body will in turn hire independent inspectors who assess whether the organic operation (and product) follows the organic standards. The inspector does not advise organic operators but will assess compliance and make recommendations to the certifying body. The certifying body will issue an organic certificate to the operator and/or identify issues of non-compliance that must be addressed. The Canadian Food Inspection Agency (CFIA) is ultimately responsible for enforcing regulations and will enter into agreements with Conformity Verification Bodies who in turn assess the practices of certification bodies on behalf of the CFIA. If the operation and its products are deemed to be following the organic standards, then the operation will receive a certificate and may claim that their product is "Organic" and attach the Canada Organic Logo (Fig. 20.3).

Fig. 20.3　The Canada Organic logo

It indicates that a product follows Canadian organic regulations. Similar logos are used by other governments around the world

Each country may have slight differences in their certification process and may require imported products to meet their own domestic standard and regulations. To make trading of organic products easier, Canada has equivalency arrangements with key markets such as the European Union and United States. In the equivalency arrangement, the organic standards and regulations are regarded as equivalent (possibly with a few exceptions), avoiding the need for exporters to get additional certification from the importing country.

The process of organic certification requires a good understanding of organic standards, a good record keeping system, and significant documentation and tracking of operations and products. Many producers in developing parts of the world have small operations, limited income, poor access to technology, and may have a low level of literacy. This means that the cost and burden of organic certification may discourage producers from becoming certified. In recognition of this, some novel approaches to certification have been established to help support such farmers. One such approach is called the Participatory Guarantee System defined as

Participatory Guarantee Systems (PGS) are locally focused quality assurance systems. They certify producers based on active participation of stakeholders and are built on a foundation of trust, social networks and knowledge exchange. (IFOAM Organics International)

Adoption of this system has allowed many small-scale organic producers from developing countries to become certified and access the organic market.

20.3.2   Organic Standards

The organic standards are "process" based standards as opposed to a "product or outcome" based standard. Thus, organic standards outline the practices that should be followed or those that are prohibited under organic production and do not make guarantees (or claims) about the healthiness, safety, or nutritional value of the products. The organic standards are not independent of other regulations. In addition to following organic standards, the certified organic operator must also adhere to all other government regulations relating to agriculture, environment, labour, animal welfare, and food safety.

Many organic standards exist around the world; some are regulated by government while others are private standards. IFOAM Organics International works with the international organic community to identify the practices which are required, permitted, and restricted under organic production. IFOAM has established a set of NORMS that provide an off-the-shelf standard for organic agriculture and guidance relating to the certification process (IFOAM Organics International, 2018). IFOAM Organics International also lists all organic standards which meet a set of criteria approved by the international community.

While organic standards vary somewhat around the world to reflect local production environments and markets, they largely have the same key components. For example, Canada's organic standards for land-based production are divided into two key documents:

1. Organic production systems: general principles and management standards (CAN/CGSB-32.310).
2. Organic production systems: Permitted Substances List (CAN/CGSB-32.310) .

Globally, few countries have standards for organic aquaculture; however, Canada has developed the "Organic Production Systems: Aquaculture—General principles, management standards and Permitted Substances Lists (CAN/CGSB-32.312)".

When asked to define organic agriculture, most people begin by listing substances and techniques that are prohibited in organic. The organic standards, however, identify ecologically based practices that support production. The Canadian management standards include sections on ①the principles of organic, ②terms and definitions, ③developing an organic plan, ④crop production, ⑤livestock production, ⑥ specific production requirements for apiculture, maple, greenhouse, sprout, and mushrooms, ⑦maintaining integrity during cleaning, preparation, and transportation, ⑧organic product composition, and ⑨procedures relating to amending the Permitted Substances List. Below a brief overview of practices that are encouraged and prohibited is presented.

20.4　Organic Production Practices

20.4.1　Organic Transition

A producer must undergo a transition period before becoming certified as organic. The transition period typically lasts 36 months since the last use of a prohibited substance. The organic standards must be followed during the transition period including maintaining all required documentation. The transition period serves several purposes: it allows the dissipation/degradation of prohibited substances; it gives the producer time to fully adopt organic management practices; and reduces opportunistic certification of producers who simply want to capture a market opportunity. The transition period can be a difficult time for producers as they do not receive a price premium during this time and yields may be lower because management practices may not have developed sufficiently for maintaining soil fertility and managing pests. The producers must learn about suitable organic equipment and practices, the standards, record keeping, and marketing of organic products. Farmers are encouraged to access resources and mentors to ease the transition process and make a gradual transition to organic production if previously using intensive conventional practices.

20.4.2　Soil Management

The organic management standards define the *practices* that are required, permitted, and prohibited. It should be remembered that the pioneers of organic agriculture believed that a production system supported by healthy soil and biodiversity would produce crops that were healthy and resilient to pest pressure, and that would be nutritious for livestock and people (Fig. 20.4). Thus, the organic standards require organic producers to support soil fertility by maintaining or enhancing soil organic matter levels, balancing nutrients, and promoting soil biology. This can be achieved by a combination of practices such as: ①crop rotation, ②applying compost or manure to fields (Fig. 20.5), ③green manure plough down, and ④supporting nutrient balance through waste recycling and supplementing nutrients from sources that were not chemically derived. Tillage should be used in a way that minimizes soil degradation and in combination with practices that improve soil quality. Greenhouses are required to grow crops in soil made up of mineral and organic matter that is supported by living organisms (Fig. 20.6). Burning of crop residues is prohibited since it depletes carbon that may have improved soil organic matter content while also affecting air quality and greenhouse gas emissions.

20.4.3   Pest Control in Crops

Control of pests (i.e. weeds, insects, diseases, rodents) is one of the greatest challenges that farmers face. Organic producers typically adopt an integrated approach to pest management which may include cultural, physical, chemical, and biological controls. Organic farmers are expected to adopt practices described as "cultural controls", practices that improve the environment or advantage of a crop or livestock while placing the pest at a disadvantage without the use of a direct control such as a pesticide or physical removal. Cultural controls start with selecting crops and livestock that are well-adapted to the growing environment and which are resistant or resilient to pest pressure. Crop rotation is used to break pest cycles by introducing different crops with different levels of resistance to an insect or disease, or which can compete more effectively with weeds. Crop density can be managed to maximize competitiveness with weeds or minimize disease pressure. Sanitization of equipment is important for preventing the introduction and spread of pests.

Cultural controls are regarded as the first line of defence against pests but they sometimes are not adequate. Organic producers can supplement cultural controls with physical controls such as mechanical weeding (Fig. 20.7), trapping, pest removal, and flaming. Biological controls involve the use of other organisms to control a pest. This may range from training and introducing livestock to eat specific weeds to the release of predators of insect pests. Mating of insect pests such as coddling moth orchards may be disrupted by introduction of pheromones which confuse the moths in trying to find a mate and reduce the likelihood of mating. Using an integrated approach to pest management can result in productive, healthy crops without the use of pesticides (Fig. 20.8).

A pesticide is typically defined as a substance or chemical that is used to kill, repel, or control an organism (i.e. plant, animal, insect, microorganism) that is considered to be a pest. Organic agriculture generally prohibits the use of synthetically derived pesticides; however, some naturally derived pesticides are permitted. As with all substances on the Permitted Substances List, pesticides using organically permitted substances undergo review beyond the pesticide product registration regulations of the government; they must also meet organic standards relating to their formulation.

Fig. 20.4　Organic market vegetable production relies on diversity not only for soil and pest management in the field but also to satisfy a wide range of consumer preferences (photo credit: A. M. Hammermeister)

This organic farm in British Columbia, Canada, produces a wide range of vegetables and cut flowers, rotating crops across their land base

Fig. 20.5　Compost is a valuable soil amendment that improves soil organic matter (humus) content and supplies nutrients (photo credit: A. M. Hammermeister)

This modern composting facility supports a commercial organic orchard in Washington state, USA

Fig. 20.6   Greenhouses such as this one in British Columbia, Canada are very important for allowing farmers to get an early start on the growing season so they can have marketable produce earlier in the season (photo credit: A. M. Hammermeister)

Fig. 20.7   Mechanical weeding equipment such as this tine weeder allows weed control after the crop has emerged without the use of herbicides (photo credit: A. M. Hammermeister)

Fig. 20.8　Well-managed organic field crops can be quite productive and have relatively few weeds as demonstrated in this soybean field in Prince Edward Island, Canada (photo credit: A. M. Hammermeister)

20.4.4　Livestock Production

The prevention of illness and injury is a priority for organic livestock producers including aquaculture. If an animal falls ill, they should be separated and treated as permitted by the organic standards. If organically permitted practices are not sufficient, the animal must receive necessary treatment to prevent suffering, even if it means the animal loses organic status. Organic livestock production standards are intended to prevent the use of chemical allopathic veterinary drugs (including antibiotics) through natural breeding, balanced nutrition, stress reduction, a healthy environment and freedom of movement and natural behaviour.

The organic standards vary among livestock as they have different requirements for feed, outdoor access, and space. Herbivores, for example, are expected to have access to pasture during the grazing season, while poultry will have access to the outdoors when weather permits. Stocking rates should consider the feed and nutritional requirements of the livestock, the production capacity of the land and environmental impact; minimum indoor and outdoor space requirements are specified in the standard. Breeds should be selected that are adapted to local conditions and the production system. Feed rations have special requirements based on the specific needs of the animal. Finally, there are guidelines for animal welfare during transport and slaughter.

20.4.5　The Permitted Substances List

A Permitted Substances List identifies the raw inputs or ingredients that can be used in organic production (as opposed to being prohibited). It includes substances that are used in production, as ingredients, as processing aids, and for cleaning/sanitation. This list is continually evolving as a better understanding of substances evolves. The addition of substances to the list follows a detailed review process that considers: ①if it complies with the principles of organic, ②if it complies with prohibitions described in the standard, ③the origin, social impacts, and ecological impacts of its sourcing and production, ④what other available alternatives there are for the substance. The Permitted Substances List included as part of the standard does not identify permitted products, trade names, or brands as these are continuously changing and their formulation often includes multiple ingredients. Thus, processed/formulated products are evaluated for their organic status based on what raw ingredients were used and the process used in manufacture.

20.4.6　Prohibited Practices and Substances

Organic agriculture prohibits several practices, techniques, and substances that may be otherwise permitted or not regulated. These include but are not limited to products and materials derived from genetic engineering, nanotechnology, livestock cloning or irradiation and other inputs including sewage waste and synthetic: crop production aids, antibiotics, growth regulators, synthetic pesticides used for storage and transport unless specified in the Permitted Substances List. These prohibitions exist because the practices or substances are not consistent with the principles of organic agriculture (as outlined above), and in taking a precautionary approach to avoid risk.

20.5　Organic Production Statistics and Market Demand

20.5.1　Production Statistics

Organic agriculture has been growing rapidly in the last few decades as described provided by the annual global statistics for organic agriculture (Willer et al., 2020). Organic agriculture is being practiced in at least 178 countries and accounts for approximately 1.5% of the global agricultural land base. Organic land has increased from 11 million ha in 1999 to over 71.5 million ha (Table 20.1). Australia has the largest organic land base followed by Argentina and China. Sixteen countries have organic practices occupying at least 10% of the agricultural land base. In some countries, organic land makes up a higher proportion of the agricultural land base such as Liechtenstein (38.5%) and Samoa (34.5%).

Table 20.1　Regional statistics of organic agricultural land, producers, and market in 2018[a,b]

Region	Land base			Producers			Market
	Organic land (ha)	Proportion of all organic land (%)	Organic proportion of total agricultural land in the world (%)	Organic producers (number)	Proportion of all organic producers (%)	10 year growth rate of organic producers (%)	Proportion of total global organic retail sales (%)
Africa	2,003,976	3	0.2	788,858	28	+53.8	0.02%
Asia	6,537,226	9	0.4	1,317,023	47	+80.5	10.4%
Europe	15,635,505	22	3.1	418,610	15	+64.3	42.2%
Latin America	8,008,581	11	1.1	227,609	8	−20.0	0.8%
North America	3,335,002	5	0.8	23,957	0.9	+42.1	45.2%
Oceania	35,999,373	50	8.6	20,859	0.7	+146.4	1.4%
World	71,514,583	100	1.5	2,796,916	100	+54.8	100

[a]Data from Schlatter et al., 2020

[b]Agricultural land presented here includes both arable and non-arable land but does not include land used for beekeeping or wild collection

The number of organic producers increased globally from 200,000 producers in 1999 to 2.8 million producers in 2018 (Table 20.1; Schlatter et al., 2020). The largest number of producers are in India (1,149,371) followed by Uganda (210,352) and Ethiopia (203,602). Over 2.4 million organic producers (86%) are from developing countries.

Permanent grassland supporting extensive livestock grazing systems makes up approximately two thirds of the organic land base, totalling 48.3 million ha. Arable crops account for over 13 million ha of production with cereals and green fodder the dominant crops (Fig. 20.9A). Fruits, olives, nuts, and coffee account for over 70% of land that is in permanent crops (Fig. 20.9B).Wild collection of plant and animal products can also form part of organic agriculture; the largest land areas under wild collection are in Finland, Zambia, and India. Organic agriculture does not just produce food and feed; over 180,000 farmers produce cotton on over 350,000 hectares of land (Barsley et al., 2020).

It is clear from the statistics provided in Table 20.1 that organic production is truly global in nature, including small- and large-scale farmers, intensive and extensive organic production, technologically advanced and constrained production, and a wide range of growing environments from tropical to arid to sub-arctic.

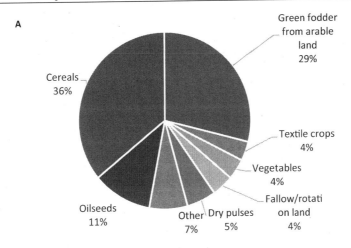

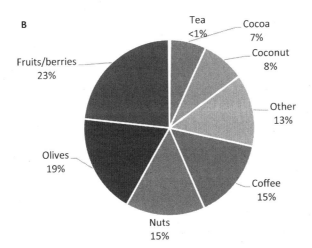

Fig. 20.9   Organic agricultural land (71.5 million ha) is comprised of permanent grassland (48.3 million ha), arable land (13.3 million ha), permanent crops (4.7 million ha), and land for which details are not available (5.2 million ha) in 2018 (Schlatter et al., 2020)

The figures above show the proportion of different crop types in (A) arable land, and (B) permanent crops

20.5.2   *Organic Markets and Consumer Motivation*

Overall, the largest markets for organic products are in North America and Europe accounting for almost 90% of the market (Table 20.1) with the largest markets in the United States, Germany, and France, however, the largest per capita consumption of organic products is in Switzerland, Denmark, and Sweden (Barsley et al., 2020). Organic markets are growing rapidly, especially in countries such as Ireland, France, Denmark, and Norway which saw increases in market growth higher than 20%; organic sales accounted for 8%–12% of total retail sales in Austria, Denmark, Luxembourg, Sweden, and Switzerland (Barsley et al., 2020). Countries with the highest value of organic exports are United States, Italy, Netherlands, and China (Barsley et al., 2020).

The organic sector would not be growing at its current pace without the support of consumer demand which supports a premium price for organic products. Numerous consumer surveys have been conducted since organic entered the marketplace to understand consumer motivations for purchasing organic products. The most common motivations related to organic purchases are free of harmful ingredients (pesticides, chemicals, additives, genetically modified content), sensory aspect (taste, flavour, colour), quality, health attribute, environment, and personal health concern (Kushwah et al., 2019). The organic industry is growing rapidly to meet this demand for organic processed goods.

20.6　Benefits and Challenges of Organic Agriculture

As mentioned above, the standards for organic agriculture are used to certify "process" rather than "outcomes". However, organic agriculture established to be distinguishable and have unique outcomes from other production systems. This is expected in order to encourage producers to adopt organic practices despite the challenges of input restrictions, pest management, maintaining yields, niche marketing, certification costs, and record keeping requirements. Organic consumers are paying premium prices for organic products with an expectation that the product or its production system has attributes unique from other products. Governments need to have justification for developing policy supporting organic agriculture.

As described earlier, the pioneers of organic agriculture intended to establish a system of production with soil health and biodiversity as the foundations for healthy crop and livestock production. The potential of adopting agroecological practices has been well-demonstrated (Fig. 20.10). The organic movement was further motivated by reducing risk from the use of synthetic pesticides. So, is organic agriculture distinguishable in its impact from the rest of agriculture?

The performance of agricultural systems is most commonly measured in terms of yield. Yields under organic production can frequently be lower than non-organic yields; however, the amount of difference depends on the crop, the growing environment, whether best practices are being compared, and whether the comparison is being made in developed or developing countries (Badgley et al., 2007; de Ponti et al., 2012; Seufert et al., 2012). In developing countries, for example, yields under organic management are expected to be higher than those under non-organic farming systems with limited access to inputs (Badgley et al., 2007).

While the Green Revolution emphasized yield increase as its primary goal, one must recognize that agriculture has many roles and impacts beyond production. Agriculture is recognized as a major contributor to biodiversity loss, pollution of water, soil degradation, greenhouse gas emissions, and health concerns. Thus, evaluating the performance of agriculture based on yield alone is not appropriate.

Fig. 20.10　This small farm in Morocco demonstrates how agroecological practices can convert a barren landscape into a diverse and productive organic farm (photo credit: A. M. Hammermeister)

Organic agriculture incorporates multiple goals within its guiding principles. When assessing the impacts of a farming system, a multifunctional approach is needed which assesses the performance based on productivity, environmental impacts, farmer well-being, and consumer benefits. When assessed in this way, organic agriculture does provide multiple benefits, particularly when assessed on a per land unit basis; the benefits are smaller, but still exist, on a per unit output basis (Reganold and Wachter, 2016; Seufert and Ramankutty, 2017).

Despite the economic, social, and environmental benefits of organic agriculture, it is difficult to cast aside the question of whether organic agriculture can feed the world. This is a complex question which requires consideration of not just production but storage losses, inequitable access to food, and trade issues. Strategies for feeding the world must seek sustainable solutions that consider climate change, food waste, and choice of land allocation to different uses. Organic agriculture can be an important model for sustainable food systems that address global food security issues if its adoption works in combination with reductions in food waste and foodcompeting feed for livestock (Muller et al., 2017).

20.7　Organic 3.0

Despite its continued growth in land base, global farmer population, and marketplace, the rate of adoption of organic agriculture has not been rapid enough to address global agri-environmental issues that some may suggest are reaching a crisis level. Leaders in the global organic community have recognized that, although organic agriculture has been a useful model for sustainable agriculture, the organic sector must grow and evolve in order to address the many challenges associated with conventional agricultural intensification and population growth (Arbenz et al.,

2016).

Organic 3.0 is a vision for the next stage in the evolution of the organic sector, where stage 1.0 was the stage of sector pioneers, and 2.0 is the current stage of regulated global trade of organic goods (Arbenz et al., 2016). The goal of Organic 3.0 is to expand the impact of organic agriculture globally by focussing on six features: ①A culture of innovation, ②Improvement towards best practice, ③Diverse ways to ensure transparency, ④Inclusion of wider sustainability interests, ⑤Empowerment from farmer to consumer, ⑥True value and cost accounting. Thus, the relevance of organic agriculture as an impactful production system that addresses global issues depends on an evolution of thinking and practice in the organic sector.

Conclusion

Organic agriculture is a system of production that consists of farmers of all types, all around the world working under a common standard. Organic agriculture can be used as a comparative model of a sustainable food system because it has standards that are globally recognized and regulated in many countries. This allows organic agriculture not only to be an identifiable "brand" for consumers to purchase, but also to help retailers understand consumer motivations in the food system, scientists to evaluate the multifunctional performance of agriculture, and policy makers to address multiple goals by supporting a single system. Organic agriculture provides an interesting platform for discussing the trade-offs between high yields and environmental impacts. The organic sector continues to evolve through Organic 3.0 as its benefits and constraints become better understood.

Chapter 21
Integrated Agroecosystem Management

E. K. Yiridoe

"To make agriculture sustainable, the grower has got to make a profit."
—Sam Farr

Abstract Agriculture is arguably the most important human activity and makes agroecosystems the most intensively managed ecosystems. Economists see farming and agriculture in a very broad context, as more than production agriculture, and extends upstream and downstream beyond farm units. Integrated farm management (IFM) is a holistic approach to managing agroecosystems in an integrated and sustainable manner. IFM provides an efficient approach to managing agricultural production systems with a goal to generate economically viable farm operations while at the same time helping to achieve socially and environmentally responsible outcomes. IFM centres on a holistic approach, with the components of the agroecosystem interacting with each other to generate synergies. A framework for inquiry into agroecosystem management needs to reflect what farmers and consumers value. The agroecosystem attributes which humans value include economic, social and environmental attributes. The profit motive is central to an economic analysis of sustainable agricultural production systems. Besides economic viability, long-term sustainability of agricultural production systems requires environmental and social responsibility and considerations. Farmers' decisions in managing agroecosystems inherently involve choices (or preferences) and trade-offs and optimization of farm resources and production risks.

Learning Objectives
After completing this chapter, students will be able to:

1. Define the following terms associated with agroecosystem management: ecosystem vs. agroecosystem; production agriculture vs. agricultural business, externalities; integrated farm management.
2. Describe the importance of "system" in the context of agroecosystem management.
3. List and describe the components or dimensions of an agroecosystem.
4. Explain why economic analysis of agroecosystem management is important.
5. Outline an economic framework for investigating agroecosystem management.
6. Describe how government policy affects agroecosystem management.

21.1　Introduction: Concepts and Definitions

To provide context and background on an economic framework for enquiry into agroecosystem management, it is important to operationalize and clarify various technical concepts and important definitions, some of which were first encountered in earlier chapters of this book. These terms and concepts include clarifying the distinction associated with: ①ecosystems versus agroecosystems and ②agriculture versus agribusiness.

Ecosystems are commonly identified as naturally occurring (unmanaged) delimited spaces or geographical areas. Examples include marshlands or watersheds, forests, lakes and deserts. Appending "agro" to "ecosystem" to obtain "agroecosystem" implies that an activity is engaged in the ecosystem by humans. The major productive human activity engaged in the ecosystem is agriculture. It is also important to highlight the meaning and relevance of "system" in "agroecosystem". In very simple terms, a system consists of parts or elements or components or dimensions that impinge on each other and interact together to accomplish an objective or set of objectives and a goal or goals. Important elements of the agroecosystem include humans and any farming activities, along with other life forms and their activities within the ecosystem (Haworth et al., 1998).

What is the distinction between farming and a farm or agricultural business? Traditional farming involves the production of food and/or fibre (in the unprocessed form). This narrow definition of agriculture represents production agriculture and includes activities such as production of poultry and animals, fruit and vegetable production (i.e., horticulture), flower production (i.e., floriculture), tree production (or viticulture), fish farming (i.e., aquaculture). However, the agriculture and agrifood sector is more than production agriculture. Economists see farming or agriculture in a broader sense as agribusiness. It encompasses various sectors of the economy upstream (input side) and downstream (output side) from a farm unit.

Stonehouse and Vander Borgh (1995) embody this approach to a description of the general structure of the agriculture and agrifood sector. Agricultural businesses span various sectors of the economy beyond the farm production unit, including:

- provision of services and inputs to agriculture (such as seeds, chemicals, machinery and equipment, veterinary services, medicines).
- processing raw materials into value-added intermediate and final food and fibre products, packaging and storage, distribution and sale to the final consumer (through grain elevators, milling, bakeries, product processing and packing plants, supermarkets and retail outlets, etc.).

Thus, the agriculture and agrifood sector, which is a part of the larger economy of a country or region, encompasses production of food and fibre and agribusiness.

Other technical terms that are important in agroecosystem management include:

Holistic approach: to farm or agroecosystem management and analysis implies a whole farm analysis or analysis across all the enterprises which constitute the farm business. Often, in a holistic analysis, it is assumed that the individual enterprises are integrated and interact among and across each other.

Externalities (associated with agricultural production): in general, externalities are uncompensated benefits and costs associated with agroecosystem management and include positive and negative externalities. An example of a positive externality relates to bees from a neighbour's apiary which fly over and pollinate a neighbouring blueberry farm. The blueberry farmer does not compensate the bee farmer for the pollination services of the bees. Similarly, an example of a negative externality relates to nitrate-N contamination of well water arising from application of excessive N fertilizer on a grain farm upstream in a watershed. The grain farmer does not compensate users of the contaminated groundwater who are harmed from drinking the well water. In the two cases described above, if the blueberry farmer fully compensates the bee farmer for the bee pollination services (positive externality) or the grain farmer fully compensates the contaminated well water users (negative externality), then the externality is said to be *internalized*.

21.2   Integrated Farm Management

The need to manage agroecosystems in a sustainable and holistic manner has prompted renewed interest in integrated farming systems and integrated farm management. Although there is no consensus on a single definition of integrated farm management (IFM), there is general agreement among scientists that the concept of IFM emerged in response to a need to balance negative externalities (i.e., uncompensated costs) from agriculture, especially negative environmental impacts of agriculture, with economic viability of the farm business. According to Hendrickson et al. (2008), integrated agricultural production systems often have multiple farm enterprises (such as crop and animal enterprises) that interact across space and time (i.e., from year to year), with the interactions often resulting in synergies among the enterprises.

The farming organization, Linking Environment and Farming (LEAF), notes that integrated farm management involves the use of the best modern technologies and traditional methods in farming, resulting in not only enriching the environment but also engaging local communities and results in site-specific and continuous improvement of the whole farm, as opposed to individual farm units (LEAF, 2017). LEAF identifies nine inter-related components that work together in a holistic manner for effective implementation of IFM: organization and planning; soil management and fertility; crop health and protection; pollution control and product management; animal husbandry; energy efficiency; water management; landscape and nature conservation and community engagement. Besides the nine components identified by LEAF, the European Initiative for Sustainable Development in Agriculture (EISA) identifies three additional components, namely climate change/air quality, human and social capital, and crop nutrition. EISA describes IFM as the "most efficient way to a productive, environmentally friendly and socially responsible agriculture" (EISA, 2012).

It was noted earlier that the various parts or components of an agroecosystem impinge on each other and interact together to achieve specific objective(s) and/or goal(s). Important production units of an integrated farming system may include crops, trees, livestock and poultry/birds.

Crops, for example, may be managed as mono-cropping systems (as in industrialized commercial agriculture) often in rotation systems or in different combinations. In other agricultural production systems (common in some developing countries), important crops may be managed as mixed or intercropping systems. Crop combinations may be selected from groups representing cereals or grains, legumes and pulses, oilseeds and forages. Similarly, trees may be incorporated into farming systems because of their economic importance in producing marketable fruits (e.g., apples and oranges), fuelwood and pellets (e.g., oak and maple) or fodder (e.g., *Acacia*). The overall goal of integrated farming systems is to balance the economic aspects of farming (i.e., productivity of farm inputs and resources) with social and environmental/ecological goals, thereby generating a farming system or agroecosystem that is sustainable and resilient over the long-term. Integrated farm management therefore inherently uses a holistic or whole farm management approach.

Examples of Integrated Farming Systems

1. Temperate Climate/Region: Mixed grain-beef operation: manure from livestock is spread on grain fields. Crops may include cereals (e.g., wheat or corn) and legume (e.g., soybeans or lentils), managed in a rotation system. Some of the grains grown on the farm are used in formulating feed rations for the farm animals. Crop residues from the multiple grains in the farm fields are ploughed back into the field.
2. Tropical Climate/Region: Aquaculture and rice production, using irrigation in a watershed. The irrigation water is important for both fish and rice production. Waste from fish production may provide nutrients when spread on the rice fields.

21.3　Economic Framework to Agroecosystem Management

21.3.1　Why Is There a Need for an Economic Framework or Approach to Agroecosystem Management?

A famous British economist, Lionel Robbins, defined economics as "the science which studies human behaviour as a relationship between ends and scarce means which have alternative uses" (Robbins, 1935). Agroecosystems are one of the most important managed ecosystems in the world.

It was noted earlier that agriculture is a major productive human activity engaged in ecosystems. It therefore makes sense that approaches for managing agroecosystems need to be based on what farmers (and other humans such as consumers) value. But what are the human values connected with managing agroecosystems?

The attributes that farmers and consumers value include not only economic and social dimensions of agroecosystems, but also services and (dis-services) associated with agroecosystems, such as contamination of water systems, greenhouse gas emissions and land and soil degradation. In modern agriculture (in both industrialized and developing countries), farming and agroecosystem management increasingly emphasizes optimization of outputs produced and conservation and long-term sustainability of agricultural production resources while at the same time minimizing negative externalities from agriculture. This implies that most farmers tend to have multiple objectives in farming and agroecosystem management.

Among the various types of ecosystems, agroecosystems are perceived as having the biggest impact on the lives of human beings because not only are agroecosystems managed to provide food and fibre, but also they have significant impacts on the quality of the environment. Management of an agricultural enterprise or business is a unique human activity because this involves managing risks associated with a need to handle and control critical inputs and resources such as farm labour, farmland, as well as capital inputs within a natural environment (i.e., the ecosystem) and economic environment (i.e., the economy). Many aspects of these environments and the dynamics within and between the environments are beyond the control of the farm manager. Examples of issues beyond the control of farmers include changing weather and climatic conditions, crop and animal disease incidence, government policy and changes in consumer behaviour and preferences towards agriculture and agrifood products and services. This lack of control over many issues which affect (agro)ecosystems makes agricultural businesses inherently riskier than non-agricultural businesses.

21.3.2 *Profit Motive in Farming*

An economic framework for agroecosystem management is centred on farming as a business (i.e., an agribusiness) in which the farmer or farm manager typically has a profit motive, along with various other non-economic objectives and goals. From an economic standpoint, the profit motive is the motivation of farm firms to generate monetary gains from their farm operations. Besides generating profit, other objectives and considerations in operating a farm business may include reducing pollution; conservation of the quantity and quality of natural resources such as soil, water and air; increasing and sustaining biodiversity, agroecosystem resilience. Critics of the profit motive in farming point to important social considerations, such as the importance of agriculture in food sovereignty and food security.

A profit motive in farming or in agroecosystem management is important for several reasons. First, a farm operation needs to generate profit (over the long-run) in order to sustain the farm, including paying for repairs of farm machinery and equipment, purchase production inputs and replace existing equipment or other inputs that are outdated or inefficient (Fig. 21.1). If the farm is not profitable, it may be difficult to obtain money to pay for such costs. A second reason why farmers need to make profit from the farm business is that most prospective investors will avoid a farm business which, over the long-run, is deemed to be not profitable. A third reason why it is important to generate profit from a farm business is that, as with other human beings, farmers need to make reasonable profit from farming in order to generate income to support their livelihoods and provide for their families (e.g., buy a car, pay school fees and health bills, pay for entertainment, save for retirement).

Fig. 21.1　Profit from a farm business is needed to purchase innovative farm equipment, such as tractors and harvesters

This picture illustrates a new labour-saving harvester for lowbush blueberries

21.3.3　*Economic Framework for Agroecosystem Management*

There is a growing need and interest to view and manage agroecosystems in a sustainable and holistic manner. As noted above, there is increasing consensus that economic viability, environmental soundness and social responsibility and acceptability are key to understanding and managing sustainable agricultural production systems or agroecosystems. Economics provides tools for analysing and optimization of whole systems such as agroecosystems, across various levels of scale in space and time.

Applied economics provides tools and models for assessing the multiple criteria and decisions and issues considered by farmers and farm managers in managing agroecosystems, as well as in evaluating the performance and sustainability of agroecosystems (Yiridoe and Weersink, 1997). *Integrated economic-biophysical optimization techniques* allow for evaluating alternative farming and management practices in agroecosystems and for assessing changes or adjustments to a farm manager's behaviour associated with introduction of new agricultural technologies, innovations or management practices (Janssen and van Ittersum, 2007). An *integrated economic-biophysical modelling* and analytical approach not only explicitly recognizes that agroecosystems have multiple dimensions, including economic and biophysical dimensions, but also captures possible interactions in space and time. This economic approach is consistent with integrated farming systems and integrated farm management.

In addition, as with all human behaviour, farmer decisions and behaviour involve choices (or preferences) and *trade-offs*. In simple terms, a trade-off is the amount of one attribute a decision-maker (such as a farmer) will give up or sacrifice (e.g., additional yield and farm profits) in order to gain specified amounts of a second attribute (e.g., less nitrate-N contamination of groundwater). Nearly all farmer decisions involve more than one attribute. This suggests a need for decision-makers to consider trade-offs. Quantifying and assessing the choices and trade-offs is critical to evaluating the sustainability of agricultural production systems and in agricultural and environmental policy analysis. Trade-offs between present and future outcomes of an agroecosystem can be used to quantify sustainability and generate quantitative measures of the sustainability of an agricultural production system (Antle et al., 1999).

Trade-off analysis of agricultural production systems or agroecosystems can be considered for several system dimensions or components at a particular time period and over time. It was noted above that besides maximizing profits, farmers may also have various other social and environmental/ecological objectives. Trade-off analysis allows for quantifying and evaluating the competing objectives in agricultural production systems.

Fig. 21.2 illustrates trade-off assessment of two states of an agroecosystem, A and B. The variables in the graph are assumed to represent optimal positive measures such that higher values of the attributes represent increasing agroecosystem sustainability. X_{1A} represents an indicator under state A (e.g., nitrate-N reduction). X_{2A} represents a second indicator under state A (e.g., crop yield) and X_{3A} say farm income. Points A, B and C are *bionomic equilibria* and represent trade-offs between indicators X_1 and X_2. Conditions A and B in the curve are considered better than situation C because C is inefficient or inferior. However, it is not so obvious if situation A (in state A) is better than B (in state A) because that depends, at least in part, on the values of the variables under consideration. Optimal values of the indicators X_1, X_2 and X_3 can be generated using integrated economic-biophysical optimization techniques. More advanced details of decision support tools and other important considerations in trade-off analysis are described elsewhere (see, for example, Antle et al., 1999; Yiridoe and Weersink, 1997).

Fig. 21.2 Assessing tradeoffs among attribute indicators (Yiridoe and Weersink, 1997)

Variables represent positive measures for which increasing values indicate increasing agroecosystem sustainability

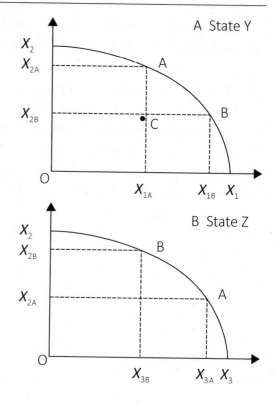

21.4 How Does Government Policy Affect Agroecosystem Management?

First, the term "policy" is clarified to provide background and context on why governments intervene in agriculture, and how this can affect agroecosystem management. Governments in both developed and developing countries set and administer policies, which are principles or actions intended to guide decisions and/or achieve particular outcomes or objectives. Agricultural policies are public policies that governments put in place to meet objectives or accomplish specific outcomes for the agriculture and agrifood sector. In this chapter, the terms "government policy" and "public policy" are used interchangeably. To understand how government or public policies affect agroecosystem management, it is important to recall the structure of the agriculture and agrifood sector described earlier. In this context, agriculture is more than just farming or production agriculture and extends beyond the farm gate, including economic sectors upstream (farm inputs and services) and post-production (or output processing and distribution).

Governments often intervene in the agriculture and agrifood sector using various policies because of (and to correct) inefficiencies in the free market system. The government policies may be put in place to help achieve various objectives, such as produce or provide public goods and services, regulate businesses or redistribute income. In addition, some external costs (and benefits) that are not fully accounted for by farm firms (and consumers) may require government intervention to internalize the externalities. For example, in Canada, governments commonly intervene in the production, marketing and distribution of agricultural goods and services. The supply management of poultry and dairy products is an example of a Canadian government agricultural policy.

In some countries such as Canada, agriculture is a shared responsibility between the federal and provincial/territorial/state governments and, therefore, affected by agri-environmental policies at the two levels of government. For example, provinces own their natural resources (although there are exceptions such as federal lands, and land owned by indigenous peoples). Canada's agricultural policy community includes federal and provincial departments of agriculture and farm and consumer groups. In other countries with non-federal systems of national government, agriculture is the responsibility of the national government. In such cases, the agricultural policy community will include regional and district departments of agriculture and farm and consumer groups. Farmers and advocates of agricultural businesses tend to be most concerned about agricultural policies and programs which affect the profitability of farm businesses and on restrictions on the right to farm.

Government agricultural and environmental policies influence many issues connected with agroecosystem management, such as: ①where farms can be located; ②what farmers can and cannot grow; ③output price received by farmers; ④how products are transported, processed and manufactured and ⑤commodity marketing and trade. This suggests that some public policies affect agroecosystem management through direct effects on farming and farm production. In addition, policies and programs such as farm subsidies and low-interest loans and guaranteed price supports are intended to raise or sustain the quantity or volume of agricultural production. Governments sometimes use *quotas* in order to increase or sustain domestic production of specific agricultural products. Agricultural policies which governments often use to alter the market or achieve particular outcomes or objectives include taxes and subsidies, price and quantity controls and tariffs (i.e., taxes paid on imports or exports). Other examples of government policies that can have effects on agriculture and agroecosystem management include government tax policy, trade policy and conservation policy. In addition, government of Canada policy on immigration is increasingly becoming important in rural and agricultural regions of the country because of constraints with finding farm labour, especially seasonal labour.

Table 21.1   Examples of government policies and their (intended) effects

Policy type[a]	Effect or intended objective or outcome
Subsidies, low-interest loans and guaranteed prices	Raise the quantity of farming output or yield
Quotas	Discourages importing particular products from other countries, thereby helping to increase domestic production
Supply management (for Canada's dairy and poultry sector)	Use of import quotas and domestic production controls to manage output supply and price levels
Immigration policy	Quantity and quality of farm labour, seasonal farmworkers
Climate change policy	Reduce greenhouse gas (GHG) emissions and mitigate changes in climate extremes

[a]These are intended as illustrative in order to highlight diversity and selected types of government policies and not-intended as (and by no means) a comprehensive or exhaustive list of policies

　　Criticisms of Canadian agricultural policy have varied over time. The criticisms generally relate to perceived neglect of important issues and interests or the failure of federal and provincial/territorial governments to optimize and balance competing political, economic and social goals of the policy (Table 21.1).

Chapter 22
Employing an Agroecological Approach to Achieve Sustainable Development Goals: A Case Study from China

Z. Si and S. Scott

*"If more of us valued food and cheer and song above hoarded
gold, it would be a merrier world."*
—*John R.R. Tolkien.*

Abstract In this chapter, using a case study approach, we examine the potential benefits of employing agroecology to achieve the United Nations Sustainable Development Goals (SDGs). As a response to food safety and environmental concerns, diverse agroecological practices have been proliferating in China in the past decade. These cases demonstrate agroecology's interdisciplinary nature in that they embody not only ecological farming cases but also social innovations (e.g., ecological farmers' cooperatives and community organizations) for ecological and healthy food provisioning. The development of agroecology in China shows how it facilitates most of the SDGs in various ways. Yet, the further development of agroecology also faces multiple challenges. This chapter reviews the contributions of agroecology to the SDGs and the barriers hindering its wider adoption. It offers lessons for other countries in terms of policy supports to enhance the capacity of producers, reduce the cost of production, and facilitate market access.

Learning Objectives

At the end of this topic, a student should be able to:

1. Discuss, with specific examples, ways in which agroecology facilitates the United Nations' Sustainable Development Goals (SDGs).
2. Describe the status of agroecology development in China.
3. List and briefly explain the barriers globally for the wider adoption of agroecology, with possible solutions to overcome those barriers.

22.1 Introduction

It is clear from the preceding chapters that implementation of the concepts of agroecology has a wide array of ecological, social, and economic benefits. In this chapter, we examine these benefits of agroecology in terms of achieving the United Nations Sustainable Development Goals (SDGs). We do this through a case study from China. Four decades after China implemented the Reform and Opening Up policy, this country's food system has undergone tremendous changes alongside its economic growth. The foremost change is reflected in the rise in food production that lifted millions out of hunger. China has been widely praised for its outstanding achievement of feeding 20% of the world's population with only 7% of the world's arable land (Lam et al., 2013; Cao et al., 2014). According to the World Food Program, the percentage of undernourishment in China was reduced from 23.9% in 1990–1992 to 9.3% in 2014–2016. China accounts for two thirds of the reduction of undernourished people in developing countries since 1990 (WFP, 2016). This achievement is obtained through a combination of institutional reforms (e.g., the decollectivization of agriculture and the establishment of the Household Responsibility System), technological innovations (e.g., hybrid seeds, machineries) (Zhang, 2019), subsidized agricultural development programs (Huang et al., 2013) and urban food system governance (Zhong et al., 2019).

China's achievements in eliminating hunger are not without cost. Over the last three decades, the damaging ecological and social consequences of boosting production have become ever more evident. China's agriculture sector faces huge environmental challenges—from overuse of fossil fuel-based chemical fertilizers, to water pollution, and soil erosion—and social and health challenges, from widespread food safety concerns (from agrochemical residues in food to adulteration of processed foods, leading to a crisis of trust), to the decline of the agricultural labor force and hollowing out of rural communities (Si, 2019; Lu et al., 2015; Wang et al., 2015b; Zolin et al., 2017; Pretty and Bharucha, 2015; Luan et al., 2013; Fang and Meng, 2013; Campbell et al., 2017; Reid et al., 2018). These environmental and social consequences result from, and are deeply embedded in, critical political economy transformations in rural China, such as the accelerated urbanization process, the shifting foci of food security policies towards promoting technologically intensive modern agriculture (see also Huang and Yang, 2017) and the concentration of land through land transfer (Day and Schneider, 2018).

These environmental and social consequences present significant challenges for China to achieve the Sustainable Development Goals (SDGs) proposed by the UN General Assembly in 2015. The sustainable development agenda consists of 17 SDGs, 126 associated targets, and 330 indicators (United Nations, 2015). As an indirect successor of the Millennium Development Goals (MDGs), the SDGs pronounced a new and universal developmental agenda that UN member states are using to frame their policies over the next 15 years. In response to the SDGs, the Chinese government issued the "National Plan on Implementation of the 2030 Agenda for Sustainable Development" (State Council, 2016). Although it specifies supports for sustainable agriculture, this plan focuses on national level programs for the state, without much consideration of opportunities for grassroots and private sector initiatives. Meanwhile, various agroecology initiatives such as ecological and organic food production, civil society organizations promoting ecological farming, ecological farmers' markets, and buying clubs, are taking root across China amid the food safety crisis of the past two decades (Scott et al., 2018).

Agroecology has been defined as "the application of ecological concepts and principles to the design and management of sustainable agroecosystems" (Gliessman, 1998). The development of agroecology offers vital opportunities and a promising approach to promote a sustainable food system in support of the SDGs (FAO, 2018). Some of the SDGs, such as SDG2 Zero Hunger, SDG3 Good Health and Wellbeing, SDG12 Responsible Consumption and Production, and SDG13 Climate Action, are directly related to the development of agroecology, meaning that the proliferation of agroecological practices could directly support these goals. Many other SDGs are indirectly connected through the relationship between food and environment, food and education, food and economic growth, as well as food and partnership building. This chapter thus focuses on the direct and the less obvious linkages between the SDGs and agroecology development, drawing from agroecology cases in China.

22.2   The Interdisciplinary Nature of Agroecology

As previously discussed, Gliessman's (1998) definition of agroecology highlights the central place of ecological principles. Yet, more diverse elements of the concept of agroecology have been developed in recent decades. Researchers suggest that a holistic understanding of agroecology encompasses multiple layers or disciplines (Dalgaard et al., 2003). Its interdisciplinary nature demands an integration of "soft" elements of agroecology in addition to "hard science" (Luo, 2018). This "soft" agroecology refers to the interaction between human and natural systems in the interrogation of agroecology. Besides the interdisciplinary nature, more recent understandings of agroecology highlight that the application of ecological concepts and principles should not only focus on food production but also reflect the entire food system (Francis et al., 2003). Agroecology should be understood as an integrative discipline that studies the ecology of the entire food system, encompassing ecological, social, and economic dimensions (Francis et al., 2003). Combining ecological principles with food system thinking (e.g., to understand the connections within the food value chain, and the multiple dimensions and scales of food issues) (Si and Scott, 2019) helps to fully appreciate the nuances of what agroecology encompasses.

Building on the recognition of both the multiple dimensions and the broad scope of agroecology, agroecology should be understood as a field that "addresses the environmental, economic and social dimensions of agri-food systems" (FAO, 2018). Because of its nature of embracing complexity (Farrelly, 2016), agroecology challenges the established notions of the conventional agrifood system that reduces the complex environmental-economic-social system to a simplified input–out system. Agroecology becomes an umbrella that covers a set of critical philosophies, knowledge, skills, and experiences to explore alternatives to the conventional agrifood system. Being rooted in a smallholder-based system and relying on ecosystem management, agroecology is a strong force of resistance to the corporate food regime that advocates for capital-intensive agriculture relying upon external inputs (Altieri, 2009; Holt-Giménez and Altieri, 2013). In contrast to the conventional agrifood system, agroecology is buttressed by small and diversified farming systems and is nourished by local communities' knowledge and innovations (Holt-Giménez and Altieri, 2013). Despite agroecology's origins among small farmers, it showcases the way in which sustainable agrifood systems operate and thus offers a path for conventional agrifood farms to transition. The fundamental differences in the principles, indicators, and practices between agroecology and the conventional agrifood system demand a close examination of how agroecology could be applied to convert the conventional system.

22.3　Agroecology Practices in China

Agroecology is recognized by many scholars and practitioners for its potential to address pressing food, fuel, and climate crises (Altieri, 2009; De Schutter, 2010; Altieri et al., 2015). Its multiple benefits are reflected in the many nontechnical innovations of agroecology in China. Both state and civil society initiatives promoting ecological food production and sustainable food consumption are emerging in China. These initiatives such as certified and uncertified ecological agriculture and civil society organizations are indeed agroecology practices because of their adoption of ecological principles in managing farms and approaching food consumption and their supports to various degrees for smallholders and local farming communities (Scott et al., 2018).

One group of agroecological initiatives are certified ecological farms. For more than two decades, the Chinese government has established different ecological agriculture standards, including hazard-free, green, and organic standards, from the lowest to the highest stringency of food production and their commitments to environmental sustainability (Scott et al., 2014). Other food quality certifications such as Geographic Indication and biodynamic certifications could also be considered a part of the broad agroecology transformation due to the various emphases on the provenance and ecological natures of food. The government, particularly the local government, has provided various supports for the expansion of these certified ecological food production in order to cope with growing domestic and international demands, the food safety concerns, and environmental awareness (Cook and Buckley, 2015; Chen et al., 2018). The organic sector is growing rapidly in the past decade, yet it also faces a crisis of consumer trust. Due to media reports of abuse of organic labels and fraud in organic certification, it is widely believed that some organic food with certification labels in the market does not meet the standards (Sternfeld, 2009; Yin et al., 2016). The over-reliance upon conventional food supply chains such as supermarkets and organic food stores reinforces the issue of distrust since, in most cases, consumers do not directly interact with farmers, and the information on how and by whom the food is produced is not always clear to consumers (Wang et al., 2015a).

The standardized third-party certification approach, such as the organic and green food certifications in China, of developing agroecology is perhaps more problematic for its exclusion or marginalization of the majority of small farmers. Small-scale farmers can rarely meet the requirements of supermarkets in terms of the levels of standardization, quality and safety standards, and the consistent supply of large volume of certain foods (Hazell et al., 2007). This exclusion is also reinforced by the high investment requirements of large-scale ecological food companies. It is also a challenge for small farmers to afford the certification cost and the demands in terms of record keeping and related paperwork for the application. Since China's revised organic certification standard implemented since 2012 (Hallman and Xu, 2012), government subsidies have only been available to producers large enough to meet the minimum requirement set by the government. This uneven playing field marginalizes small farmers in terms of conducting certified ecological food production and earning a price premium.

Overcoming the distrust of third-party certification schemes among Chinese consumers who purchase organic food requires rebuilding trust through alternative systems of food provisioning. It is within this context that uncertified ecological agriculture initiatives are proliferating in China, such as community supported agriculture (CSA) farms and other uncertified ecological farms established either by individual agricultural entrepreneurs who call themselves "new farmers", or by traditional peasant farmers (Scott et al., 2018). CSAs in China emerged as an alternative response to the growing food safety concerns and environmental awareness among the "middle class" in Chinese cities (Shi et al., 2011). Aside from its innovative scheme of reconnecting farmers with consumers, one of the key features of CSAs is the adoption of ecological principles in farming. Diverse and mixed ecological farming approaches such as traditional farming skills, organic farming, natural farming, biodynamic farming, and permaculture were found on these CSA farms (Schumilas, 2014; Cook and Buckley, 2015). In most cases, ecological farmers in China are constantly conducting experiments of applying ecological principles to farming.

The application of agroecology in China is not limited to CSAs; it takes diverse forms. Cook and Buckley (2015) , for example, documented the diverse organizational structure of ecological farms, particularly the important role of farmers' cooperatives in coordinating production, harvesting, and marketing. Amongst the cases, sustainable agriculture and ecological agriculture are interpreted and practiced in multiple ways, linking local communities, external actors, and the market. They demonstrate multiple pathways of transitioning towards agroecological systems.

The proliferation of agroecology initiatives in China would not be possible without the various social organizations promoting agroecology (Cook and Buckley, 2015; Scott et al., 2018). Both national and local non-profit organizations or consumer organizations are playing critical roles in carrying forward the agroecological transition. To name a few, the New Rural Reconstruction Network, the CSA Coalition based in Beijing, Partnerships for Community Development based in Hong Kong, and NurtureLand based in Guangzhou, alongside many others, spearheaded the development of ecological agriculture. Being the key driver of many initiatives, these organizations have connected farmers with concerned consumers and academics and have even brought agroecological transition of agriculture to the local government's agenda, such as the government supports for CSA development in Guiyang, Shanghai, Beijing, Changzhou, and Lankao (Zhang and

Lin, 2017). It is also increasingly obvious that the development of agroecology in China is generating serious policy attention at the highest level, as reflected in the connection between the proliferation of agroecology and China's agenda to approach the UN SDGs.

22.4　Connections Between Agroecology and the UN Sustainable Development Goals

In order to understand the potential of agroecology in facilitating the SDGs, we need to closely examine the targets listed under each of the SDGs and select the most relevant targets. The development of agroecology is directly linked to many of the SDGs, including SDG1 (No poverty), SDG2 (zero hunger), SDG3 (good health and wellbeing), SGD6 (clean water and sanitation), SDG8 (decent work and economic growth), SDG11 (sustainable cities and communities), SDG12 (responsible consumption and production), SDG13 (climate action), SDG15 (life on land) and SDG17 (partnership for the goals). Furthermore, the agroecological transition of the food system is also related to other SDGs indirectly. These close connections make the comprehensive adoption of agroecology a powerful entry point and instrument to facilitate the SDGs. Table 22.1 exemplifies how the contemporary development of agroecology in China is contributing to the SDG's. It highlights the most relevant discourses in the targets under each goal. The connection is reflected in agroecology's capacity in enhancing those specific targets. In the following section, we use three agroecology cases in China to showcase their potential for facilitating the SDG's.

22.5　Case Studies

22.5.1　Case 1: Nested Market and SDG 1 No Poverty

An inspirational initiative that connects villagers in Sanggang village, Hebei province, with consumers in Beijing is known as a "nested market". Nested markets, according to Jan Douwe van der Ploeg (2012), refer to new food markets where "producers and consumers are linked through specific networks and commonly shared frames of reference". Through rebuilding consumer and producer connections, nested markets emerged out of rural development processes as a constructed response to the control of food empires in the food supply chain. In the Chinese context, the nested market was initiated by consumers who were increasingly skeptical about the quality and safety of food channeling through the modern food supply chains. Ecological production and the associated quality of food are thus a critical feature of a product that nurtures the new network. Besides the economic value of food, social and environmental values are also embedded in the multilevel exchanges among the people within the nested market.

Table 22.1 Linking agroecology experiences in China with the SDG's

SDGs	Explanations	Relevant targets	Connections with agroecology in China
SDG1: No poverty	End poverty in all its forms everywhere	1.5 By 2030, build the *resilience* of the poor and those in vulnerable situations and reduce their exposure and vulnerability to climate-related extreme events and other economic, social, and environmental shocks and disasters	• Consumer organizations buying ecologically produced food directly from farmers contribute to poverty reduction by introducing alternative market opportunities to enhance the economic resilience of the poor • Farmers' cooperatives provide economic opportunities for smallholders marginalized in the market economy
SDG2: Zero hunger	End hunger, achieve food security and improved nutrition and promote sustainable agriculture	2.1 access to *safe, nutritious, and sufficient food* all year round; 2.3 By 2030, double the agricultural productivity and *incomes of smallscale food* producers...through opportunities for *value addition* 2.4 By 2030, ensure *sustainable food production systems* and implement resilient agricultural practices 2.5 By 2020, maintain the *genetic diversity* of seeds, cultivated plants and farmed and domesticated animals and their related wild species	• Ecological agriculture, such as CSA farms, facilitates the access to safe and nutritious food for urban residents, increases the income of smallholders, promotes sustainable food production systems and practices, contributes to biodiversity, and encourages investment in agriculture • Participatory plant breeding programs contribute to the conservation of the genetic diversity of seeds
SDG3: Good health and well-being	Ensure healthy lives and promote Well-being for all at all ages	3.9 By 2030, substantially reduce the number of deaths and illnesses from *hazardous chemicals and air, water, and soil pollution and contamination*	• Ecological and organic farming not only reduces the exposure of farmers and farm workers to hazardous chemicals but also reduces consumers' intake of chemical residues through reduction of water and soil pollution and chemical sprays and therefore contributes to public health
SDG4: Quality education	Ensure inclusive and equitable quality education and promote lifelong learning opportunities for all	4.7 By 2030, ensure that all learners acquire the *knowledge and skills* needed to promote sustainable development, including, among others, through *education* for sustainable development and sustainable lifestyles, ...appreciation of cultural diversity and culture's contribution to sustainable development	• Agroecology constitutes a critical element of sustainable development education • Farmer-to-farmer informal education provides practical, relevant information for enhancing agroecological practices

(continued)

SDGs	Explanations	Relevant targets	Connections with agroecology in China
SDG4: Quality education	Ensure inclusive and equitable quality education and promote lifelong learning opportunities for all	4.A Build and upgrade *education facilities* that are child, disability and gender sensitive and provide safe, non-violent, inclusive, and effective learning environments for all	• Traditional farming knowledge in China exemplifies culture's potential contribution to sustainable development • CSAs and school gardens also provide inclusive and effective learning venues for children and their parents
SDG5: Gender equality	Achieve gender equality and empower all women and girls	5.5 Ensure *women's* full and effective participation and equal opportunities for leadership at all levels of decisionmaking in political, economic, and public life 5.A undertake reforms to *give women equal rights to economic resources*, as well as access to ownership and *control over land* and other forms of property, financial services, inheritance, and natural resources, in accordance with national laws	• CSAs and other ecological farms valorize women's, especially 'left-behind' women's, roles in supporting the family and the development of the community • Ecological farming and farmers' cooperatives as emerging livelihood opportunities enhance women's access to and control over land and other natural resources and thus improves gender equality
SDG6: Clean water and sanitation	Ensure availability and sustainable management of water and sanitation for all	6.3 By 2030, improve water quality by reducing pollution, *eliminating dumping and minimizing release of hazardous chemicals and materials*, halving the proportion of untreated wastewater, and substantially increasing recycling and safe reuse globally 6.4 By 2030, substantially *increase water-use efficiency* across all sectors and ensure sustainable withdrawals and supply of freshwater to address water scarcity and substantially reduce the number of people suffering from water scarcity 6.6 By 2020, protect and *restore water-related ecosystems*, including mountains, forests, wetlands, rivers, aquifers, and lakes	• Agroecology reduces the usage of agrochemicals and thus contributes to water pollution reduction • It promotes water conservation practices through building soil organic matter and use of mulching • Ecological farming, such as circular agriculture, encourages water recycling, uses localadaptive seed varieties and technologies and thus increases water-use efficiency • Ecological farmers work with nature in a non-exploitive way and thus contributes to restore water-related ecosystems
SDG7: Affordable and clean energy	Ensure access to affordable, reliable, sustainable, and modern energy for all		• Agroecology reduces reliance on fossil fuel by avoiding the use of synthetic, fossil fuel-based chemicals and uses ecological technologies for pest control and fertility building

(continued)

SDGs	Explanations	Relevant targets	Connections with agroecology in China
SDG8: Decent work and economic growth	Promote sustained, inclusive, and sustainable economic growth, full and productive employment, and decent work for all	8.3 Promote development-oriented policies that support productive activities, *decent job creation, entrepreneurship, creativity and innovation,* and encourage the formalization and growth of micro-, smalland medium-sized enterprises 8.4 improve progressively, through 2030, global *resource efficiency* in consumption and production and endeavor to decouple economic growth from environmental degradation 8.5 By 2030, achieve full and *productive employment* and decent work for all women and men 8.6 By 2020, substantially reduce the proportion of *youth* not in employment, education, or training 8.8 Protect labor rights and promote *safe and secure working environments* for all workers, including migrant workers, in particular women migrants, and those in precarious employment 8.9 By 2030, devise and implement policies to promote *sustainable tourism* that creates jobs and promotes local culture and products	• Agroecology provides various opportunities for entrepreneurs, creates jobs for farmers and returning farmers, contributes to innovation in terms of farming approaches, marketing strategies, and organizational structures • Agroecology improves resource efficiency in food production through the management and recycle of resources • Ecological farmers' cooperatives and CSAs provides decent employment opportunities for both young entrepreneurs and marginalized women farmers, enabling them to stay in rural areas • Ecological production environment is safer for all workers compared to industrial farms • Eco-villages and eco-cities where agroecology plays a significant role are often popular tourists sites and agritourism promotes the sale of local products
SDG9: Industry, innovation and infrastructure	Build resilient infrastructure, promote inclusive and sustainable industrialization, and foster innovation	9.2 Promote *inclusive and sustainable industrialization*	• The rapidly developing organic food industry in China offers an example of sustainable industrialization • Opportunities for value-adding through food processing • The development of agroecology in China is an inclusive development process because it is a joint effort of various actors and involves both academic and grassroots innovations
SDG10: Reduced inequalities	Reduce inequality within and among countries	10.2 By 2030, empower and promote the *social, economic, and political inclusion* of all, irrespective of age, sex, disability, race, ethnicity, origin, religion, or economic or other status	• ecological farmers' cooperatives empower small farmers by ensuring economic opportunities and their access to resources

(continued)

SDGs	Explanations	Relevant targets	Connections with agroecology in China
SDG11: Sustainable cities and communities	Make cities and human settlements inclusive, safe, resilient, and sustainable	11.6 By 2030, reduce the adverse per capita environmental impact of cities, including by paying special attention to air quality and municipal and other *waste management* 11.A support positive economic, social, and environmental *links between urban, per-urban, and rural areas* by strengthening national and regional development planning	• Urban and periurban agriculture such as community gardens can convert organic waste into fertilizer and thus reduce waste • CSAs in China strengthen urban and rural connections through communitybuilding activities • Eco-agritourism also support positive urban–rural linkages
SDG12: Responsible consumption and production	Ensure sustainable consumption and production patterns	12.2 By 2030, achieve the *sustainable management and efficient use of natural resources* 12.3 By 2030, halve per capita global *food waste* at the retail and consumer levels and reduce food losses along production and supply chains, including post-harvest losses 12.5 By 2030, substantially *reduce waste* generation through prevention, reduction, recycling, and reuse 12.8 By 2030, ensure that people everywhere have the relevant *information and awareness for sustainable development* and lifestyles in harmony with nature 12.A support developing countries to strengthen their scientific and technological *capacity* to move towards more sustainable patterns of consumption and production	• Sustainable resource management is promoted through traditional ecological farming approaches such as intercropping, rotation, rice-duck and rice-fish systems promotes sustainable and efficient use of sunlight, soil, and water resources • Agroecology at small-scale farms and circular agriculture reduce post-harvest food waste and turn waste into resources • Ecological farms through their marketing, on-farm activities, social media platforms, newsletters, and daily conversations with members are vanguards of increasing public awareness for sustainable development and consumption • Agroecology is a cutting edge research field in China for strengthening capacities to advance sustainable production
SDG13: Climate action	Take urgent action to combat climate change and its impacts	13.1 strengthen *resilience and adaptive* capacity to climaterelated hazards and natural disasters in all countries 13.2 Integrate climate change measures into national *policies, strategies, and planning* 13.3 Improve education, awareness-raising and human and institutional *capacity* on climate change mitigation, adaptation, impact reduction, and early warning	• Agroecology enhances the resilience and adaptive capacity to climate change of the agriculture sector and advocates policy changes that integrate climate change concerns • Agroecology as a rapidly developing discipline in Chinese universities greatly improves knowledge and awareness for climate change

(continued)

SDGs	Explanations	Relevant targets	Connections with agroecology in China
SDG14: Life below water	Conserve and sustainably use the oceans, seas, and marine resources for sustainable development		• Ecological and organic farming prevents run-off from overuse of nitrogen fertilizers and therefore ocean acidification and eutrophication
SDG15: Life on land	Protect, restore, and promote sustainable use of terrestrial ecosystems, sustainably manage forests, combat desertification, and halt and reverse land degradation and halt biodiversity loss	15.3 By 2030, combat desertification, restore *degraded land and soil*, including land affected by desertification, drought, and floods, and strive to achieve a land degradation-neutral world 15.4 By 2030, ensure the conservation of *mountain ecosystems*, including their biodiversity, in order to enhance their capacity to provide benefits that are essential for sustainable development 15.6 Promote fair and equitable sharing of the benefits arising from the utilization of *genetic resources* and promote appropriate access to such resources, as internationally agreed 15.9 By 2020, integrate ecosystem and biodiversity values into *national and local planning*, development processes, *poverty reduction* strategies, and accounts	• Agroecology such as small-scale ecological farming in remote areas in China enhances sustainable management of soil and local ecosystems, through building of soil organic matter, etc. • Agroecology through participatory plant breeding and conservation projects protects the genetic resources of traditional crop varieties and animal breeds and encourages the equitable sharing of the benefits • With China's on-going poverty alleviation campaign, agroecology is increasingly recognized for its potential to lift small farmers out of poverty. Some ecological farmers' cooperatives, for example, were incorporated into the poverty reduction project by local governments
SDG16: Peace, justice and strong institutions	Promote peaceful and inclusive societies for sustainable development, provide access to justice for all, and build effective, accountable, and inclusive institutions at all levels	16.7 ensure responsive, inclusive, participatory, and representative *decision- making* at all levels	• Agroecology promotes the sovereignty of farmers, gives voice to small farmers in the agriculture sector • Cooperatives and participatory plant breeding promote inclusive and participatory development
SDG17: Partnership for the Goals	Strengthen the means of implementation and revitalize the global partnership for sustainable development	17.14 Enhance *policy coherence* for sustainable development 17.15 Respect each country's policy space and leadership to establish and implement policies for *poverty eradication* and sustainable development *Multi-stakeholder partnerships* 17.17 Encourage and promote effective public, public–private, and civil society *partnerships*	• Agroecology promotes food system thinking that coordinates policy making across different governance sectors • Ecological agriculture opens a promising space for poverty eradication and sustainable development through the partnership among farmers, entrepreneurs, consumers, researchers, and local governments

The diverse outcomes of the experiment of the nested market in Sanggang village showcased how agroecology development could contribute to poverty reduction alongside many other social and environmental goals. Since 2007, employees of China Agricultural University have connected with Sanggang villagers to buy fresh produce, eggs, and meat produced with limited use of agrochemicals. Trust between villagers and their customers was established through village visits and direct conversations. By 2017, about 200 Beijing customers were ordering food every month from over 130 households. The exchange between villagers and their urban customers is not limited to food but also embodies multidimensional information, a sense of care, knowledge, and skills. The strong interest from urban consumers incentivizes villagers to reduce the use of chemicals in farming and improve the environment in the village to attract visitors. With the support of the local government, member households of the Sanggang village nested market established a farmers' cooperative. This modest income from selling food enabled the elders to have a decent livelihood in the village. The assistance from their customers, including establishing and training villagers in managing the online ordering platform, and providing suggestions for improving customer services, built villagers' capacity for self-governing the nested market. The social connections in the village were revitalized through the project. As van der Ploeg et al. (2012) argues, rural development is about "new networks that link the rural and the urban". The nested market, albeit quite small, is an innovative network that prevents villagers from becoming impoverished.

22.5.2　Case 2: Ecological Farmers' Cooperatives and SDG 8 Decent Work and Economic Growth

Farmers' cooperatives, also commonly known as farmers' professional cooperatives, have been growing rapidly in China since the 1990s with government support. The growth has been greatly accelerated after the "Farmers' Cooperative Law" was enforced in 2007 (Song et al., 2014). Rural China has undergone a period of "cooperative fever" (Hu et al., 2017). However, the policies for promoting farmers' cooperatives have mainly focused on the economic dimensions of cooperatives (i.e., food production and marketing), with much less emphasis on the social and environmental dimensions (Song et al., 2014). Overlooking the critical roles of smallholders in the organization of farmers' cooperatives, most registered cooperatives turned out to be fake "shell cooperatives" or private agrifood companies (Hu et al., 2017).

Although most so-called farmers' cooperatives are not authentic, some ecological farmers' cooperatives that apply ecological principles to food production have played critical roles in local economic growth. An interesting case is the Organic Farmers' Cooperative of Dai village, Jurong city in Jiangsu province. Daizhuang used to be a village suffering from poverty. Conventional farming with agrochemicals prevailed in the village until 2001 when the retired agricultural technician Zhao Yafu helped the villagers to convert to ecological and organic agriculture. The first few years were spent on educating farmers about organic farming, and to convince them of the viability of organic farming in terms of yields and profits. In 2006, the Daizhuang Organic Farmers' Cooperative was founded to produce and market organic rice. One year after that, all the 600 farming households in the village joined the cooperative. By pooling about 3000 mu (~494 acres) of land together, each household became a shareholder of the cooperative. About 80% of the cooperative income was paid back to farmer shareholders, while the other 20% was retained as collective savings. As non-farming job opportunities are increasing in nearby cities, many farmers left the farm for other jobs but still kept their land in the cooperative to receive a dividend. In 2017, about 200 farming households participated in organic farming and each of them generated an annual income of 100,000 CNY (~14,800 USD), higher than the income of many urban households in Jurong city. The development of the Organic Farmers' Cooperative in Dai village lifted the poor villagers out of poverty. While many villagers worked as migrant workers in the city, they could still be shareholders through contributing their land to the cooperative. The agroecological transition also greatly improved the environment of the village, preparing it well for developing agritourism and sustainable multifunctional agriculture (Xinhua News, 2018).

Farmers' cooperatives have been a critical and even dominant organizational approach for agroecological production in China, as demonstrated in Cook and Buckley's (2015) book. Cases from across China confirm that cooperatives provide critical farming inputs such as seeds, organic fertilizer, and bio-pesticides, as well as training, and technical guidance for practicing agroecology to their members. Farmers' cooperatives also address technical and marketing constraints facing individual small-scale farmers, enable them to meet requirements of getting government subsidies, and organize eco-tourism. Cooperatives provide decent work to small farmers and enabled economic growth in remote rural areas. It also greatly contributes to promoting "sustainable tourism that creates jobs" and "local culture and products", which are all targets of SDG 8.

22.5.3   Case 3: Community Organizations of CSAs in China and SDG 17 Partnership for the Goals

The development of CSAs in China began in the late 2000s. By providing ecologically produced food to urban residents, CSAs are direct responses to the widespread food safety anxiety. In the early days, CSA farms were located across Chinese major cities without much connection. Nevertheless, civil society organizations such as the Green Ground based in Beijing and Partnership for Community Development (PCD) based in Hong Kong began to link these CSAs together. The Green Ground, albeit small in the early days of CSA development, connected CSA farms with consumers by helping farmers to sell their products to consumers in Beijing. PCD on the other hand facilitated some of the earliest CSA farms in southwest China. One of the most well-known CSA farms in China, the Little Donkey Farm in Beijing, also functioned as a knowledge hub for training interns who were motivated to establish their own CSA farms. Perhaps a more influential civil organization in China involved in agroecological development among smallholders is the New Rural Reconstruction Movement (NRRM). As a national network initiated by researchers from Renmin University, the NRRM has played a prominent role in building solidarity among CSA farms and other agroecological initiatives across the country (Si and Scott, 2016). Through its annual CSA symposium and the CSA coalition network, the NRRM made great achievements in spreading the knowledge and skills of sustainable development of agriculture and beyond. The knowledge mobilization is based on a well-communicated network of ecological farmers, food activists, researchers, ecological agricultural entrepreneurs, consumer groups, civil society organizations, policy makers, and international organizations. With many more civil society organizations emerged across the country, the development of agroecology is increasingly enriched with diverse perspectives, ideas, and experiences (Scott et al., 2018).

The proliferation of social organizations in sustainable agriculture in China demonstrates how the development of agroecology significantly encourages and promotes "effective public-private and civil society partnerships", a key objective of the SDG17. As the influence of CSAs is growing rapidly, the government has recognized its potential in facilitating governmental agendas in sustainable development, such as improving food safety conditions, developing sustainable agriculture, reducing rural poverty, and stimulating rural economic growth. Therefore, recent CSA conferences have witnessed an increasing presence of the state. For example, the tenth National CSA Conference held in Chengdu in 2018 was strongly supported by the government. Government officials from the Ministry of Agriculture and Rural Affairs, the Ministry of Ecology and Environment, and the Chengdu Government attended the conference. The CSA community in China is gradually transitioning from a minor network to a network with mainstream and official recognition. The development of agroecology opened an opportunity for the many civil society organizations to cultivate their relationship with the government and facilitate public and civil society partnerships. These partnerships are critical to China's sustainable development in the long run.

22.6　Moving Agroecology Forward for the SDGs

Many of the connections between agroecology and the SDGs have not yet been recognized. There are numerous barriers to moving agroecology forward around the world. The IPES-Food (2016) report, for example, identified eight "lock-in" factors that keep industrial agriculture in place and constrains the expansion of agroecology. Most of these factors also apply to the Chinese situation, although some of them may unfold in different ways. These barriers relate to the bias in food policies, the lack of understandings of consumers, the increasing power of agribusiness, and international trade orders. Besides these overarching structural constraints, four additional barriers specific to sustainable food systems in China exacerbate these lock-ins and create additional obstacles to agroecological transition.

- The first of these China-specific barriers relates to the *lack of consistent and committed government support.* Despite China's agriculture policies gradually incorporating more sustainability concerns recently—such as the "ecological civilization" directive—the transition towards agroecology systems in China still has a long way to go. The first obstacle is the unequal access to government supports for agroecological production and research. Not all ecological farms can benefit from financial or material government support. These supports often favor large-scale producers and exclude small ecological farmers.

- The second barrier relates to *the rejection and loss of traditional agricultural knowledge through rapid industrialization.* Chinese farmers have historically been renowned for their sophisticated traditional knowledge and technologies of ecological farming (King, 1911). However, for the past several decades, there has been promotion of agricultural practices that include increased use of agrochemicals; this has instilled in farmers a tendency towards more intensive agriculture. Much of the legacy of knowledge of traditional ecological farming approaches has already been lost. Many ecological farmers who have tried to recruit traditional farmers to work on their farm found that farmers did not know how to farm without using pesticides and chemical fertilizers and in most cases, they were sceptical of the ecological approaches they were told to follow. At the same time, many farmers have taken the path of highly commercialized agriculture, dependent on high yields and mechanization. Changing from such an industrial approach to agriculture into a more ecological approach is therefore a challenge.

- The *issue of trust* is a third fundamental issue for China. Chinese consumers' lack of trust in ecological products imposes a major challenge on most ecological farms—addressing all kinds of critiques from their customers. The cost of establishing and maintaining trust is prohibitively high such that many smallscale farmers found it too challenging to find customers. The lack of trust is not only an issue for consumers but also an obstacle for ecological farmers. According to a recent survey conducted by Chen Weiping from Renmin University, a large proportion (~80%) of the products being sold as ecological or bio-pesticides were in fact contaminated with synthetic chemicals. In fact it may even be worse than that; in 2015, the Ministry of agriculture discovered that 96% of the bio-pesticides sampled did not meet quality standards (Wang, 2016). They found that many products were incorrectly labeled and did not indicate the true active ingredients. This makes Chinese ecological farmers reluctant to accept the free bio-fertilizer distributed by some local governments.

- Fourth, the *economic viability of ecological agriculture in China* is a challenge to farmers wanting to move through the transition period from conventional farming to an ecological one. Even after the transition period, many farmers are finding it difficult to make a profit, despite the fact that at the consumer level prices are higher for organic or ecological foodstuffs. This is likely due to ①the enormous marketing challenges in an environment of low consumer trust; and ②the increasing costs of production (e.g., access to land, labor, farming inputs) without access to government supports. Adding to the problem of consumers' distrust is the increased occurrences of products on the market which are sold as organic but are in fact fraudulent (China Daily, 2018).

The cases of agroecology in China demonstrate that most of the SDGs could be facilitated through the development of agroecology. This study offers lessons to other countries aiming to employ agroecology to achieve the SDGs. The cases show the necessity of comprehensively reviewing the policy environment in order to identify and remove policy barriers for the wider adoption of agroecology in conventional agrifood systems. This implies the need to prioritize ecology and farmers in developing agricultural policies and programs. It also implies more direct supports for ecological agriculture and its food value chain to enhance the capacity of producers, reduce the costs of production, and facilitate market access. What is more, these supports should not exclude small-scale producers. More training programs are needed to revive the traditional farming knowledge and disseminate agroecology principles and practices to farmers. To enhance trust, policies should aim to create a more enabling environment for diverse private and civil society initiatives. Public programs for food education (food literacy) are also urgently needed for rebuilding consumer trust, promoting sustainable and healthy eating, and enhancing consumers' appreciation of the social and environmental benefits of agroecology.

References

Aaen SM, Helgesen KO, Bakke MJ, et al. (2015) Drug resistance in sea lice: a threat to salmonid aquaculture. Trends Parasitol, 31: 72–81

Adhikari K, Hartemink A (2016) Linking soils to ecosystem services—a global review. Geoderma, 262: 101–111

Akbari H, Davis S, Dorsano S, et al. (1992) Cooling Our Communities: A Guidebook on Tree Planting and Light-colored Surfacing. US Environmental Protection Agency, Washington, DC

Alberta Agriculture, Food and Rural Development (2005) Manure composting manual. https://www1.agric.gov.ab.ca/$department/deptdocs.nsf/all/agdex8875/$file/400_27-1.pdf? Accessed 25 May 2021

Alkorta I, Albizu I, Amezaga I, et al. (2004) Climbing a ladder: a step-by-step approach to understanding the concept of agroecosystem health. Rev Environ Health 19: 141–159

Alkorta I, Albizu I, Garbisu C (2003) Biodiversity and agroecosystems. Biodivers Conserv 12: 2521–2522

Altieri MA (1989) Agroecology: a new research and development paradigm for world agriculture. Agric Ecosyst Environ, 27: 37–46

Altieri MA (1995) Agroecology: the Science of Sustainable Agriculture, 2nd ed. Intermediate Technology Publications, London

Altieri MA (1999) The ecological role of biodiversity in agroecosystems. Agric Ecosyst Environ, 74: 19–31

Altieri MA (2005) Agroecology: principles and strategies for designing sustainable farming systems. Third World Network, Ecological Agricultural and Food Security. http://www.fao.org/agroecology/database/detail/en/c/893012/. Accessed 25 May 2021

Altieri MA (2009) Agroecology, small farms, and food sovereignty. Mon Rev 61(3). Available via DIALOG https://monthlyreview.org/2009/07/01/agroecology-small-farms-and-foodsovereignty/

Altieri MA, Nicholls CI, Henao A, et al. (2015) Agroecology and the design of climate changeresilient farming systems. Agron Sustain Dev, 35(3): 869–890

American Forests (2000) CityGreen: Calculating the Value of Nature. American Forests, Washington, DC, pp 59

Andrich MA, Imberger J (2013) The effect of land clearing on rainfall and fresh water resources in Western Australia: a multi-functional sustainability analysis. Int J Sustain Dev World Ecol, 20: 549–563

Antle J, Stoorvogel M, Crissman J, et al. (1999) Tradeoff assessment as a quantitative approach to agricultural/environmental policy analysis. In: Proceedings of the Third International Symposium on Systems Approaches for Agricultural Development (SAAD), Lima, Peru, pp 810

Arbenz M, Gould D, Stopes C (2016) Organic 3.0 for Truly Sustainable Farming and Consumption, 2nd ed. IFOAM Organics International and the Sustainable Organic Agriculture Action Network. https://www.ifoam.bio/sites/default/files/2020-3/summary organic3.0 web 1.pdf.Accessed 27 May 2021

Asbjornsen H, Hernandez-Santana V, Liebman M, et al. (2014) Targeting perennial vegetation in agricultural landscapes for enhancing ecosystem services. Renewable Agric Food Syst, 29(2): 1–25

Averill C, Waring B (2018) Nitrogen limitation of decomposition and decay: how can it occur? Glob Chang Biol, 24: 1417–1427

Ayer NW, Tyedmers PH (2009) Assessing alternative aquaculture technologies: life cycle assessment of salmonid culture systems in Canada. J Clean Prod, 1: 362–373

Badgley C, Moghtader J, Quintero E, et al. (2007) Organic agriculture and the global food supply. Renewable Agric Food Syst, 22: 86–108

Barsley L, Truscott L, Tan E, et al. (2020) Organic cotton. In: Willer H, Schlatter B, Trávníček J, et al. (eds) The World of Organic Agriculture. Statistics and Emerging Trends 2020. Research Institute of Organic Agriculture (FiBL), Frick, and IFOAM-Organics International, Bonn, pp 132–136

Beheshti FM, Parrish CC, Wells J, et al. (2018) Minimizing marine ingredients in diets of farmed Atlantic salmon

(*Salmo salar*): effects on growth performance and muscle lipid and fatty acid composition. PLoS One, 13(9): e0198538

Bell C, Gaydos L, Rowantree R (1993) Measuring the urban growth and changing vegetative structure of Sacramento, California using Landsat digital data. In: Proceedings of the 1993 GIS Symposium. Vancouver, BC, pp 1193–1199

Béné C, Barange M, Subasinghe R, et al. (2015) Feeding 9 billion by 2050-putting fish back on the menu. Food Sec, 7: 261–274

Bünemann EK, Bongiorno G, Bai Z, et al. (2018) Soil quality—a critical review. Soil Biol Biochem, 120: 105–125

Bodensteiner P (1999) Stormwater Changes Direction. Builder Magazine, Washington, DC, pp 48–54

Boiteau G, Pelletier Y, Misener GC, et al. (1994) Development and evaluation of a plastic trench barrier for protection of potato from walking adult Colorado potato beetles (Coleoptera: Chrysomelidae). J Econ Entomol, 87: 1325–1331

Boody G, Vondracek B, Andow DA, et al. (2005) Multifunctional agriculture in the United States. Bioscience, 55: 27–38

Bornstein RD (1968) Observations of the urban heat island effect in New York city. J Appl Meteorol, 7: 575–582

Bostock J, McAndrew B, Richards R, et al. (2010) Aquaculture: global status and trends. Philos Trans R Soc B, 365(1554): 2897–2912

Brooks S (2012) Best Policy Guidance for the Integration of Biodiversity and Ecosystem Services in Standards. The Secretariat of the Convention on Biological Diversity, Montréal, pp 1

Brown L (1984) A Worldwatch Institute Report on Progress toward a Sustainable Society. The Worldwatch Institute, Worthington, DC

Brugmann J (1996) Planning for sustainability at the local government level. Environ Impact Assess Rev 16(4–6): 363–379

Caldwell CD (1996) IN100: Agroecology Course Resource Manual. NSCC, Dartmouth, pp 215

Campbell BM, Beare DJ, Bennett EM, et al. (2017) Agriculture production as a major driver of the earth system exceeding planetary boundaries. Ecol Soc, 22(4): 8

Canadian Forest Service (1995) Silvicultural Terms in Canada. Natural Resources Canada, Ottawa, ON

Cao L, Naylor R, Henriksson P, et al. (2015) China's aquaculture and the world's wild fisheries. Science, 347: 133–135

Cao S, Lv Y, Zheng H, et al. (2014) Challenges facing China's unbalanced urbanization strategy. Land Use Policy, 39: 412–415

Capstaff NM, Miller AJ (2018) Improving the yield and nutritional quality of forage crops. Front Plant Sci, 9: 535

Carson R (1962) Silent Spring. Houghton Mifflin Company, Boston

Chaparro-Africano AM (2019) Toward generating sustainability indicators for agroecological markets. Agroecol Sustain Food Syst, 43: 40–66

Chen A, Scott S, Si Z (2018) Top-down initiatives: state support for ecological and organic agriculture in China. In: Scott S, Si Z, Schumilas T, et al. (eds) Organic Food and Farming in China: Top-down and Bottom-up Ecological Initiatives. Routledge, London, pp 38–59

China Daily (2018) Fake organic products a blight on market that must be rooted out. Available via DIALOG http://www.chinadaily.com.cn/a/201805/15/WS5afa1d7da3103f6866ee8559.html. Accessed 10 Jan 2020

Ciais P, Sabine C, Bala G, et al. (2013) Carbon and other biogeochemical cycles. In: Stocker TF (ed) Climate Change 2013: the Physical Science Basis. Contribution of Working Group I to the Fifth Assessment Report of the Intergovernmental Panel on Climate Change. Cambridge University Press, Cambridge

CIMMYT (2019) Maize Research. https://www.cimmyt.org/work/maize-research/. Accessed 5 Mar 2020

Collinge S (1996) Ecological consequences of habitat fragmentation: implications for landscape architecture and planning. Landsc Urban Plan, 36: 59–77

Colombo SM, Parrish CC, Wijekoon MPA (2018) Optimizing long chain-polyunsaturated fatty acid synthesis in salmonids by balancing dietary inputs. PLoS ONE, 13(10): e0205347

Colville D (1998) Developing a GIS-based tool for defining natural landscape units. Linking protected areas with working landscape conserving biodiversity. In: Proceedings of the Third International Conference on Science and Management of Protected Areas. SAMPAA, Wolfville, NS, pp 668–676

COMEST (World Commission on the Ethics of Scientific Knowledge and Technology) (2005) The Precautionary Principle. United Nations Educational, Scientific, and Cultural Organization, Paris

Conway GR (1985) Agroecosystem analysis. Agric Adm, 20: 31–55

Cook S, Buckley L (2015) Multiple Pathways: Case Studies of Sustainable Agriculture in China. IIED, London

Costanza R (1992) Toward an operational definition of ecosystem health. In: Costanza R, Norton B, Haskell B (eds) New Goals for Environmental Management. Island Press, Washington, DC

Costanza R (2012) Ecosystem health and ecological engineering. Ecol Eng, 45: 24–29

Costanza R, d'Arge R, de Groot RS, et al. (1997) The value of the world's ecosystem services and natural capital. Nature, 387: 253–260

Cox DN, Reynolds J, Mela DJ, et al. (1996) Vegetables and fruits: barriers and opportunities for greater consumption. Nutr Food Sci, 5: 44–47

Crowther TW, Todd-brown KEO, Rowe CW, et al. (2016) Quantifying global soil carbon losses in response to warming. Nature, 540: 104–108

Daily G, Alexander S, Ehrlich P, et al. (1997) Ecosystem services: benefits supplied to human societies by natural ecosystems. Issues in Ecology, 1: 1–6

Daily GC (1997) Nature's Services: Societal Dependence on Natural Ecosystems. Island Press, Washington, pp 392

Daley CA, Abbott A, Doyle PS, et al. (2010) A review of fatty acid profiles and antioxidant content in grass-fed and grain-fed beef. Nutr J, 9: 10

Dalgaard T, Hutchings NJ, Porter JR (2003) Agroecology, scaling and interdisciplinarity. Agric Ecosyst Environ, 100(1–3): 39–51

David W (1996) Community tree planting: early survival and carbon sequestering potential. J Arboric, 22(5): 222–228

Day AF, Schneider M (2018) The end of alternatives? Capitalist transformation, rural activism and the politics of possibility in China. J Peasant Stud, 45(7): 1221–1246

de Groot RS, Wilson MA, Boumans RMJ (2002) A typology for the classification, description and valuation of ecosystem functions, goods and services. Ecol Econ, 41: 393–408

de Ponti T, Rijk B, van Ittersum MK (2012) The crop yield gap between organic and conventional agriculture. Agric Syst, 108: 1–9

De Schutter O (2010) Report submitted by the special rapporteur on the right to food. Human rights council 16th session. In: United Nations General Assembly, New York

de Villiers M (2000) Water. Stoddart Publishing Co. Limited, Toronto, ON

Deng L, Zhu GY, Tang ZS, et al. (2016) Global patterns of the effects of land-use changes on soil carbon stocks. Global Ecol Conserv, 5: 127–138

Dhaliwal MS (2017) Classification of Vegetable Crops, 3rd ed. Kalyani Publishers, pp 12–17

dos Santos NZ, Dieckow J, Bayer C, et al. (2011) Forages, cover crops and related shoot and root additions in no-till rotations to C sequestration in a subtropical ferralsol. Soil Tillage Res, 111: 208–218

Duchemin M, Hogue R (2009) Reduction in agricultural non-point source pollution in the first year following establishment of an integrated grass/tree filter strip system in southern Quebec (Canada). Agric Ecosyst Environ, 131: 85–97

Dunster J (1998) The role of arborists in providing wildlife habitat and landscape linages throughout the urban forest. J Arboric, 24(3): 160–167

Dwyer JF, Nowak D, Nobel MH, et al. (1999) Connecting people with ecosystems in the 21st century: an assessment of our nation's urban forests. USDA Forest Service Technical Report PNW-GTR-490, Pacific Northwest Research Station

Dyck S, Peschke G (1995) Grundlagen der Hydrologie, 3rd ed. Aufl. Verlag für Bauwesen, Berlin

Edwards P, Zhang W, Belton B, et al. (2019) Misunderstandings, myths and mantras in aquaculture: its contribution to world food supplies has been systematically over-reported. Mar Policy, 106: 103547

Ehrlich PR, Mooney HA (1983) Extinction, substitution, and ecosystem services. Bioscience, 33: 248–254

Elergsma A (2015) Grazing increases the unsaturated fatty acid concentration of milk from grassfed cows: a review of the contributing factors, challenges and future perspectives. Eur J Lipid Sci Technol, 117: 1345–1369

Elmendorf WF, Luloff AE (1999) Using ecosystem based and traditional land use planning to conserve greenspace. J Arboric, 25(5): 264–280

European Commission (2006) Disclosure of origin in IPR applications: options and perspectives of users and providers of genetic resources. https://www.ecologic.eu/sites/files/download/projekte/ 1800-1849/1802/wp8_final_report.pdf. Accessed 5 Dec 2019

European Initiative for Sustainable Development in Agriculture (EISA) (2012) Sustainable agriculture: what is it all about? Available via DIALOG http://sustainable-agriculture.org/wpcontent/uploads/2012/07/BrochureEISA_ECPA_web.pdf. Accessed 3 Oct 2019

Falkenmark M, Rockström J (2006) The new blue and green water paradigm: breaking new ground for water resources planning and management. J Water Resour Plan Manage, 132: 129–132

Fang L, Meng J (2013) Application of chemical fertilizer on grain yield in China analysis of contribution rate: based on principal component regression C-D production function model and its empirical study. Chin Agric Sci Bull, 29(17): 156–160

FAO (1988) Sustainable Agricultural Production: Implications for International Agricultural Research. TAC Secretariat, Roma

FAO (2007) Agriculture and water scarcity: a programmatic approach to water use efficiency and agricultural productivity. In: Work Paper of Committee on Agriculture Twentieth Session

FAO (2015a) Soil functions. FAO, Rome. http://www.fao.org/resources/infographics/infographicsdetails/en/c/284478/. Accessed 15 Jan 2020

FAO (2015b) Soils and biodiversity. FAO, Rome. http://www.fao.org/soils-2015/news/newsdetail/en/c/281917/. Accessed 30 Jan 2020

FAO (2018) Food Outlook—Biannual Report on Global Food Markets—November 2018. FAO, Rome, pp 104

FAO (2018) Scaling up agroecology initiatives: transforming food and agricultural systems in support of the SDGs. Rome: United Nations Food and Agriculture Organization. Available via DIALOG http://www.fao.org/3/I9049EN/i9049en.pdf

FAO (2018) The state of world fisheries and aquaculture 2018: meeting the sustainable development goals. Food & Agriculture Org, Rome

FAO (2018) World Food and Agriculture Statistical Pocketbook. FAO, Rome

FAO (2019) Agroforestry. FAO. http://www.fao.org/forestry/agroforestry/en/. Accessed 6 Sep 2019

FAO (2019) State of food security and nutrition in the world. Available via DIALOG http://www. fao.org/state-of-food-security-nutrition. Accessed 15 Mar 2020

FAO (2019) The state of the world's biodiversity the state of the world's biodiversity for food and agriculture. FAO, Rome. http://www. fao. org/3/CA3129EN/CA 3129. PDF. Accessed 5 Jun 2021.

FAO (2020) Cereal supply and demand brief. http://www.fao.org/worldfoodsituation/csdb/en/. Accessed 5 Mar 2020

FAO (2020) The state of the world fisheries and aquaculture 2020: meeting the sustainable development goals. Food & Agriculture Org, Rome

FAO (2020) World food situation. Available via DIALOG http://www.fao.org/worldfoodsituation/csdb/en/. Accessed 02 Mar 2020

FAO, ITPS (2015) Status of the World's soil resources (SWSR) —Main report. In: Food and Agriculture Organization of the United Nations and Intergovernmental Technical Panel on Soils. Rome. http://www.fao.org/3/bc590e/bc590e.pdf.Accessed 25 May 2021

Fiedler AK, Landis DA (2007) Attractiveness of Michigan native plants to arthropod natural enemies and herbivores. Environ Entomol, 36: 751–765

Foley JA, DeFries R, Asner GP, et al. (2005) Global consequences of land use. Science, 309: 570–574

Fonte SJ, Six J (2010) Earthworms and litter management contributions to ecosystem services in a tropical agroforestry system. Ecol Appl, 20: 1061–1073

Ford AE, Graham H, White PC (2015) Integrating human and ecosystem health through ecosystem services frameworks. EcoHealth, 12: 660–671

Francini-Filho RB, Asp NE, Siegle E, et al. (2018) Perspectives on the great amazon reef: extension, biodiversity, and threats. Front Mar, Sci 5: 142

Francis C, Lieblein G, Gliessman S, et al. (2003) Agroecology: the ecology of food systems. J Sustain Agric, 22(3): 99–118

Fraser D (2009) Animal behaviour, animal welfare and the scientific study of affect. Appl Anim Behav Sci, 118: 108–111

Friesen L (1998) Impacts of Urbanization on plant and bird communities on forest ecosystems. For Chron, 74(6): 855–859

Gerla PJ (2007) Estimating the effect of cropland to prairie conversion on peak storm run-off. Restor Ecol, 15: 720–730

Gliessman SR (1998) Agroecology: Ecological Processes in Sustainable Agriculture. Ann Arbor Press, Chelsea

Gliessman SR (2005) Agroecology: the Ecology of Sustainable Food System. CRC Press, London

Globe Newswire (2019) World vegetable market analysis, forecast, size, trends, and insights. Available via DIALOG "World-Vegetable-Market Analysis, Forecast, Size, Trends and Insights". researchandmarkets.com. Accessed 19 Sep 2019

Glover KA, Solberg MF, McGinnity P, et al. (2017) Half a century of genetic interaction between farmed and wild Atlantic salmon: status of knowledge and unanswered questions. Fish and Fisheries, 18: 890–927

Gómez-Cortés P, Juárez M, de la Fuente MA (2018) Milk fatty acids and potential health benefits: an update vision. Trends Food Sci Technol, 81: 1–9

Goodland R (1992) The case that the world has reached its limits: more precisely that the current throughput growth in the global economy cannot be sustained. Popul Environ, 13: 3

Government of Canada (2019) Safe food for Canadians regulations. SOR/2017–108. Current to 29 July 2019. Minister of Justice, Government of Canada. Available via DIALOG http://lawslois. justice.gc.ca

Grant J, Simone M, Daggett T (2019) Long-term studies of lobster abundance at a salmon aquaculture site, eastern Canada. Can J Fish Aquat Sci, 76: 1096–1102

Green Infrastructure Ontario Coalition (2019) About green infrastructure. Available via DIALOG https://greeninfr astructureontario.org/. Accessed 22 Nov 2019

Gregory PJ, Ingram JSI (2000) Global change and food and forest production: future scientific challenges. Agric Ecosyst Environ, 82: 3–14

Grey D, Garrick D, Blackmore D, et al. (2013) Water security in one blue planet: twenty-first century policy challenges for science. Philos Trans Roy Soc, 371: 20120406

Group CsBSR (1998) China's Biodiversity Status Research Report, China's Biodiversity Status Research Group. China's Environment Press, Beijing

Grumbine RE (1994) What is ecosystem management? Conserv Biol, 8: 27–38

Guillen J, Natale F, Carvalho N, et al. (2019) Global seafood consumption footprint. Ambio, 48: 111–122

Hallman A, Xu F (2012) Organic certification issues update from east China. Global Agricultural Information Network report. Available via DIALOG https://gain.fas.usda.gov/Recent%20GAIN%20Publications/Briefing% 20on%20the%20Organic%20Certification%20Issues%20in%20East%20China_Shanghai%20ATO_China%20-% 20Peoples%20Republic%2of_4-16-2012.pdf

Harris RW, Clark J, Matheney N (1999) Arboriculture: Integrated Management of Landscape Trees, Shrubs and Vines. Prentice Hall Publishing, Upper Saddle River, NJ, pp 612

Hasse JE, Lathrop RG (2003) Land resource impact indicators of urban sprawl. J Appl Geogr, 23 (2003): 159–175

Haworth L, Brunk C, Jennex D, et al. (1998) A dual-perspective model of agroecosystem health: system functions and system goals. J Agric Environ Ethics, 10(2): 127–152

Hazell P, Poulton C, Steve S, et al. (2007) The Future of Small Farms for Poverty Reduction and Growth. International Food Policy Research Institute, Washington, D.C

Heckman J (2005) A history of organic farming: transitions from sir Albert Howard's war in the soil to USDA national organic program. Renewable Agric Food Syst, 21: 143–150

Heisler GM, Grimmond S, Grant R, et al. (1994) Investigation of the influence of Chicago's urban forests on wind and air temperature within residential neighborhoods. In: Chicago's Urban Forest Ecosystem: Results of the Chicago Urban Forest Climate Project. USDA Forestry Service General Technical Report NE-186. Northeast Forest Experiment Station, Randor, PA, pp 19–40

Hellin J, William LA, Cherrett I (1999) The Quezungual system: an indigenous agroforestry system from western Honduras. Agro Syst, 46: 229–237

Hendrickson JR, Hanson JD, Tanaka DL, et al. (2008) Principles of integrated agricultural systems: introduction to processes and definition. Renewable Agric Food Syst, 23(4): 265–271

Henriquez AF, Molina-Murillo SA (2017) Methodological proposal to quantify and to compensate the agroecosystem services generated by the good agricultural practices of small-farmers. Ecosistemas, 26: 89–102

Hicks C, Cohen PJ, Graham NAJ, et al. (2019) Harnessing global fisheries to tackle micronutrient deficiencies. Nature, 574: 95–98

Holt-Giménez E, Altieri MA (2013) Agroecology, food sovereignty, and the new green revolution. Agroecol Sustain Food Syst, 37(1): 90–102

Howard SA (1943) An Agricultural Testament. Oxford University Press, Inc, London

Hsieh W (2000) Rural entrepreneurs in the Canton Delta region: shift of silk to sugarcane in the 1930s. Paper presented at the 5th annual meeting of the Association for Asian Studies, San Diego

Hu XJ, Xiong YC, Li YJ, et al. (2014) Integrated water resources management and water users' associations in the arid region of Northwest China: a case study of farmers' perceptions. J Environ Manage, 145: 162–169

Hu Z, Zhang QF, Donaldson JA (2017) Farmers' cooperatives in China: a typology of fraud and failure. China J, 78: 1–24

Huang J, Wang X, Rozelle S (2013) The subsidization of farming households in China's agriculture. Food Policy, 41: 124–132

Huang J, Yang G (2017) Understanding recent challenges and new food policy in China. Glob Food Sec, 12: 119–126

Hubacek K, Sun L (2005) Changes in China and their effects on water use. J Ind Ecol, 9: 187–200

Huber B, Schmid O, Batlogg V (2018) Standards and regulations: organic regulations update. In: Willer H, Lernoud J (eds) The World of Organic Agriculture: Statistics and Emerging Trends 2018. Research Institute of Organic Agric. (FiBL), Frick, & IFOAM – Organics International, Bonn, pp 151–159

IFOAM Organics International (2018) The IFOAM NORMS for organic production and processing version 2014. Available via DIALOG https://www.ifoam.bio/sites/default/files/ifoam_norms_july_2014_t.pdf. Accessed 26 Oct 2019

IFOAM Organics International (2019a) Definition of organic agriculture. Available via DIALOG https://www.ifoam.bio/en/organic-landmarks/definition-organic-agriculture. Accessed 26 Oct 2019

IFOAM Organics International (2019b) Principles of organic agriculture. Available via DIALOG https://www.ifoam.bio/sites/default/files/poa_english_web.pdf. Accessed 26 Oct 2019

IFOAM Organics International (2019c) History. Available via DIALOG https://www.ifoam.bio/en/about-us/history. Accessed 26 Oct 2019

IFOAM Organics International (2019d) Participatory guarantee system. Available via DIALOG https://www.ifoam.bio/en/search?find=participatory+guarantee+system. Accessed 26 Oct 2019

Index Mundi (2020) Agricultural production, supply, and distribution. https://www.indexmundi.com/agriculture/?commodity. Accessed 15 Mar 2020

IPES-Food (2016) From Uniformity to Diversity: a Paradigm Shift from Industrial Agriculture to Diversified Agroecological Systems. International Panel of Experts on Sustainable Food Systems, Paris

Jackson LE, Pascual U, Hodgkin T (2007) Utilizing and conserving agrobiodiversity in agricultural landscapes. Agric Ecosyst Environ, 121: 196–210

Janssen S, van Ittersum M(2007) Assessing farm innovations and responses to policies: a review of bio-economic farm models. Agric Syst, 94: 622–636

Jones-Walters L, Mulder I (2009) Valuing nature: the economics of biodiversity. J Nat Conserv, 17: 245–247

Joseph B, Jini D (2013) Antidiabetic effects of *Momordica charantia* (bitter melon) and its medicinal potency. Asian Pac J Trop Dis, 3(2): 93–102

Kang P, Chen W, Hou Y, et al. (2018) Linking ecosystem services and ecosystem health to ecological risk assessment: a case study of the Beijing-Tianjin-Hebei urban agglomeration. Sci Total Environ, 636: 1442–1454

Kearney SP, Fonte SJ, Garcia E, et al. (2019) Evaluating ecosystem service trade-offs and synergies from slash-and-mulch agroforestry systems in El Salvador. Ecological Indicators, 105: 264–278

Kearns C (2010) Conservation of biodiversity. Nat Edu Knowl, 1: 7

Keith AM, Schmidt O, McMahon BJ (2016) Soil stewardship as a nexus between ecosystem services and one health.

Ecosyst Serv, 17: 40–42

King FH (1911) Farmers of Forty Centuries. Or Permanent Agriculture in China, Korea and Japan. Rodale Press, Inc., Pennsylvania

Kipinski B, Hanson C, Lomax J, et al. (2013) Reducing food loss and waste. World Resources Institute working paper. https://files.wri.org/d8/s3fs-public/reducing_food_loss_and_waste. pdf. Accessed 25 May 2021

Klaus IS, McPherson E, Simpson J (1998) Air pollution uptake by Sacramento's urban forest. J Arboric, 24(4): 224–234

Konijnendijk C, Annerstedt M, Nielsen AB, et al. (2013) Benefits of Urban Parks: a systematic review—a report for IFPRA (The International Federation of Parks and Recreation Administration)

Kotar J (2000) Ecologically based forest management on private lands. University of Minnesota Extension Service. Available via DIALOG http://www.na.fs.fed.us/spfo/pubs/misc/ecoforest/toc.htm. Accessed 12 Nov 2019

Kushwah SA, Dhir M, Gupta B (2019) Determinants of organic food consumption: a systematic literature review on motives and barriers. Appetite 143: 1. https://doi.org/10.1016/j.appet.2019. 104402

Lajeunesse D (1994) The use of geographical information systems (GIS) for the ecological management of urban woodlots. In: Proceedings of the First Canadian Urban Forest Conference. Canadian Forestry Association, Ottawa, ON, pp 126–130

Lal R (2016) Soil health and carbon management. Food Energy Sec, 5: 212–222

Lam HM, Remais J, Fung MC, et al. (2013) Food supply and food safety issues in China. Lancet, 381 (9882): 2044–2053

Lashomb JH, Ng YS (1984) Colonization by the colorado potato beetle, *Leptinotarsa decemlineata* (Coleoptera: Chrysomelidae) in rotated and non-rotated potato fields. Environ Entomol, 13: 1352–1356

Lavallee JM, Soong JL, Cotrufo MF (2020) Conceptualizing soil organic matter into particulate and mineral-associated forms to address global change in the 21st century. Glob Chang Biol, 26: 261–273

Li D, Liu J, Chen H, et al. (2018) Forage grass cultivation increases soil organic carbon and nitrogen pools in a karst region, Southwest China. Land Degrad Dev, 29(12): 4397–4404

Lin BB (2011) Resilience in agriculture through crop diversification: adaptive management for environmental change. Bioscience, 61: 183–193

Linking Environment and Farming (LEAF) (2017) Integrated farm management: an overview. Available via DIALOG https://leafuk.org/farming/integrated-farm-management. Accessed 20 Sep 2019

Lipper L, Thornton P, Campbell BM, et al. (2014) Climate-smart agriculture for food security. Nat Clim Chang, 4: 1068–1072

Liu J, Diamond J (2005) China's environment in a globalizing world. Nature, 435: 1179–1186

Liu Y, Rosten TW, Henriksen K, et al. (2016) Comparative economic performance and carbon footprint of two farming models for producing Atlantic salmon (*Salmo salar*): land-based closed containment system in freshwater and open net pen in seawater. Aquac Eng, 71: 1–12

Liu Y, Song W (2019) Modelling crop yield, water consumption, and water use efficiency for sustainable agroecosystem management. J Clean Prod, 253: 119940

Lloyd S, Wong T, Chesterfield C (2002) Water Sensitive Urban Design—a Stormwater Management Perspective. In: An Industry Report of Cooperative Research Centre for Catchment Hydrology & Department of Civil Engineering, Monash University, Melbourne

Lu Y, Jenkins A, Ferrier RC, et al. (2015) Addressing China's grand challenge of achieving food security while ensuring environmental sustainability. Sci Adv, 1(1): 1–5

Luan J, Qiu H, Jing Y, et al. (2013) Decomposition of factors contributed to the increase of China's chemical fertilizer use and projections for future fertilizer use in China. J Nat Res, 28 (11): 1869–1878

Luo S (2018) Agroecological Rice Production in China: Restoring Biological Interactions. FAO, Rome. http://www.fao.org/3/ca0100en/CA0100EN.pdf.Accessed 27 May 2021

Luo SM, Yan F, Chen F (1987) Agroecology. Hunan Science & Technology Press, Changsha

Lynch DH (2014) Sustaining soil organic carbon, soil quality and soil health in organic field crop management systems. In: Martin RC, MacRae R (eds) Managing Energy, Nutrients and Pests in Organic Field Crops. CRC Press, Boca Raton, pp 107–132

Lynch DH (2015) Nutrient cycling and soil health in organic cropping systems—importance of management strategies and soil resilience. Sustain Agric Res, 4: 76–84

Lynch DH (2019a) Soil is the key to our planet's history and future. https://theconversation.com/soil-is-the-key-to-our-planets-history-and-future-116330. Accessed 28 Oct 2019

Lynch DH (2019b) How soil carbon can help tackle climate change. https://theconversation.com/ how-soil-carbon-can-help-tackle-climate-change-116039. Accessed 28 Oct 2019

Ma BL, Liang BC, Biswas DK, et al. (2012) The carbon footprint of maize production as affected by nitrogen fertilizer and maize-legume rotations. Nutr Cycl Agroecosyst, 94: 15–31

Mahon JR, Miller RW (2003) Identifying high-value greenspace prior to land development. J Arboric, 29: 25–33

Makovnikova J, Kobza J, Palka B, et al. (2016) An approach to mapping the potential of cultural Agroecosystem services. Soil Water Res, 11: 44–52

Mann C, Lynch DH, Fillmore S, et al. (2019) Relationships between field management, soil health and microbial community composition. Appl Soil Ecol, 144: 12–21

Mao D, Wang Z, Wu J (2018) China's wetlands loss to urban expansion. Land Degrad Dev, 29: 2644–2657

McIntyre NE, Knowles-Yánez K, Hope D (2000) Urban ecology as an interdisciplinary field: differences in the use of "urban" between the social and natural sciences. Urban Ecosyst, 4: 5–24

McPherson EG (1994) Energy Saving Potential of Trees in Chicago. USDA Forestry Service General Technical Report NE-186. Northeast Forest Experiment Station, Randor, PA, pp 95–114

McPherson EG (1998) Atmospheric carbon dioxide reduction by Sacramento's urban forest. J Arboric, 24(4): 215–223

McPherson EG, Rowantree R (1993) The energy conservation potential of urban tree planting. J Arboric, 19(6): 321–331

MEA (Millenium Ecosystem Assessment) (2005) Ecosystems and Human Well-being, Synthesis. MEA, Washington, DC

Meadows DH (1972) Limits to Growth. Signet Press Seattle, WA

Mendoza RG (2003) The natural history of maize. Gale Encyclopedia of Food & Culture. http://www.academia.edu/350222/Maize The Natural History of Maize. Accessed 26 May 2021

Milestad R, Ahnstrom J, Bjorklund J (2011) Essential multiple functions of farms in rural communities and landscapes. Renewable Agric Food Syst, 26: 137–148

Milewski I, Loucks RH, Fisher B, et al. (2018) Sea-cage aquaculture impacts market and berried lobster (*Homarus americanus*) catches. Mar Ecol Prog Ser, 598: 85–97

Millennium Ecosystem Assessment Program (2005) Ecosystems and Human Well-being. Island Press, Washington, DC

Miller RW (1997) Urban Forestry: Planning and Managing Urban Greenspaces. Prentice-Hall, Englewood Cliffs, NJ, pp 399

Milligan RDA, Raedeke KJ (1995) Wildlife habitat design in urban forest landscapes. In: Bradley GA (ed) (1995) Urban Forest Landscapes: Integrating Multidisciplinary Perspectives. University of Washington Press, Seattle, WA, pp 139–149

Milner M, Kung KJS, Wyman JA, et al. (1992) Enhancing overwintering mortality of colorado potato beetle (Coleopetra: Chrysomelidae) by manipulating the temperature of its habitat. J Econ Entomol, 85: 1701–1708

Minasny B, Malone BP, McBratney AB, et al. (2017) Soil carbon 4 per mille. Geoderma, 292: 59–86

Montoya PAT, Valencia NM (2018) Collective action and association of heterogeneities in agroecological farmers' markets: Asoproorganicos (Cali, Colombia). Revista Colombiana De Sociologia, 41: 83–101

Moonen AC, Barberi P (2008) Functional biodiversity: an agroecosystem approach. Agric Ecosyst Environ, 127: 7–21

Mora C, Caldwell IR, Caldwell JM, et al. (2015) Suitable days for plant growth disappear under projected climate change: potential human and biotic vulnerability. PLoS Biol, 13(6): e1002167

Morton A, Routledge R (2016) Risk and precaution: salmon farming. Mar Policy, 74: 205–212

Motha RP, Baier W (2005) Impacts of present and future climate change and climate variability on agriculture in the temperate regions: North America. Climatic Change, 70: 137–164

Muller A, Schader C, Scialabba N, et al. (2017) Strategies for feeding the world more sustainably with organic agriculture. Nat Commun 8: 1290

Myers N, Mittermeier RA, Mittermeier CG (2000) Biodiversity hotspots for conservation priorities. Nature, 403:

853–858

Nahuelhual L, Defeo O, Vergara X, et al. (2019) Is there a blue transition underway? Fish and Fisheries, 20: 584–595

Naylor RL, Hardy RW, Bureau DP, et al. (2009) Feeding aquaculture in an era of finite resources. PNAS, 106: 15103–15110

Nelson KL, Lynch DH, Boiteau G (2009) Assessment of changes in soil health throughout organic potato rotation sequences. Agric Ecosyst Environ, 131: 220–228

Neori A, Chopin T, Troell M, et al. (2004) Integrated aquaculture: rationale, evolution and state of the art emphasizing seaweed biofiltration in modern mariculture. Aquaculture, 231: 361–391

Nichols RB (1999) The destruction of wildlife habitat by suburban sprawl and the mitigating effects of land use planning. A report in the Proceedings of the 1999 Northeastern Recreation Symposium. General Technical Report NE-269. North Eastern Research Station. United States Department of Agriculture, Randor, PA

NOAA (2019) Arctic report card. Available via https://arctic.noaa.gov/Report-Card/Report-Card- 2019/ArtMID/ 7916/ArticleID/835/Surface-Air-Temperature. Accessed 15 Feb 2020

Novikova A, Rocchi L, Vitunskiene V (2015) Consumers' willingness to pay for agroecosystem services in Lithuania: first results from a choice experiment pilot survey. In: Raupeliene A (ed) Proceedings of 2017 International Scientific Conference on Rural Development. Aleksandras Stulginskis University, Lithuania, pp 16

Nowak D (1993) Historical vegetation change in Oakland and its implications for urban forest management. J Arboric, 19(3): 173–177

Nowak D (1994a) Air pollution removal by Chicago's urban forest. In: Chicago's Urban Forest Ecosystem: Results of the Chicago Urban Forest Climate Project. USDA Forestry Service General Technical Report NE-186. Northeast Forest Experiment Station, Randor, PA, pp 63–81

Nowak D (1994b) Atmospheric carbon dioxide reduction by Chicago's urban forest. In: Chicago's Urban Forest Ecosystem: Results of the Chicago Urban Forest Climate Project. USDA Forestry Service General Technical Report NE-186. Northeast Forest Experiment Station, Randor, PA, pp 63–81

Nowak D (1994c) Urban forest structure: the state of Chicago's urban forest. In: Chicago's Urban Forest Ecosystem: Results of the Chicago Urban Forest Climate Project. USDA Forestry Service General Technical Report NE-186. Northeast Forest Experiment Station, Randor, PA, pp 3–18

NRC (Natural Resources Canada) (n.d.) Soil. https://www.nrcan.gc.ca/our-natural-resources/forests-forestry/sustainable-forest-management/conservation-protection-canadas/soil/13205. Accessed 19 Jan 2020

Obiero K, Meulenbroek P, Drexler S, et al. (2019) The contribution of fish to food and nutrition security in Eastern Africa: emerging trends and future outlooks. Sustainability, 11: 1636

Odum EP (1975) Ecology, the Link between the Natural and the Social Sciences. Holt, Rinehart and Winston, New York

Odum EP, Barrett GW (1977) Fundamentals of Ecology. Cengage Publishing, South Melbourne

Osterholz WR, Liebman M, Castellano MJ (2018) Can soil nitrogen dynamics explain the yield benefit of crop diversification? Field Crop Res, 219: 33–42

Oxford Dictionary Editorial Board (2019) Agriculture. Oxford Dictionary Editorial Board, Oxford

Pankhurst CE, Doube BM, Gupta VVSR (1997) Biological indicators of soil health. In: Pankhurst CE, Doube BM, Gupta VVSR (eds) Biological Indicators of Soil Quality. CAB International, Wallingford, New York, pp 421

Paoletti MG, Pimentel D, Stinner BR, et al. (1992) Agroecosystem biodiversity—matching production and conservation biology. Agric Ecosyst Environ, 40: 3–23

Pauleit S, Duhme F (2000) GIS assessment of Munich's urban forest structure for urban planning. J Arboric, 26(3): 133–141

Pauly D, Zeller D (2016) Catch reconstructions reveal that global marine fisheries catches are higher than reported and declining. Nat Commun, 7: 10244

Paustian K, Lehmann J, Ogle S, et al. (2016) Climate-smart soils. Nature, 532: 49–57

Pavlikakis GE, Tsihrintzis VA (2000) Ecosystem management: a review of a new concept and methodology. Water Resour Manag, 14: 257–283

Pedigo LP, Rice ME (2009) Entomology and Pest Management, 6th ed. Waveland Press Inc, Long Grove, Il

Peterson EE, Cunningham SA, Thomas M, et al. (2017) An assessment framework for measuring agroecosystem health. Ecol Indic, 79: 265–275

Plaster E (2009) Soil Science and Management, 5th ed. Delmar Cengage Learning, New York

Porter J, Costanza R, Sandhu H, et al. (2009) The value of producing food, energy, and ecosystem services within an agro-ecosystem. Ambio, 38: 66–93

Postma-Blaauw MB, de Goede RG, Bloem J, et al. (2010) Soil biota community structure and abundance under agricultural intensification and extensification. Ecology, 9: 460–473

Pretty J (2008) Agricultural sustainability: concepts, principles, and evidence. Philos Trans R Soc, 363: 447–465

Pretty J, Bharucha ZP (2015) Integrated pest management for sustainable intensification of agriculture in Asia and Africa. Insects, 6: 152–182

Pretty J, Brett C, Gee D, et al. (2001) Policy challenges and priorities for internalising the externalities of modern agriculture. J Environ Plann Manage, 44: 263–283

Qingyang G (2017) Integrating soft and hard infrastructures for inclusive development. J Infrastructure Policy Dev, 1(1): 1

Rapport DJ (1989) What constitutes ecosystem health? Perspect Biol Med, 33: 120–132

Rapport DJ, McMichael AJ, Costanza R (1998) Assessing ecosystem health. Trends Ecol Evol, 13: 397–402

Rauba E (2017) Services of surface water ecosystems in relations to water usage for irrigation of agricultural land. Ekonomia I Srodowisko-Econ Environ, 4: 143–155

Reganold JP, Wachter JM (2016) Organic agriculture in the twenty-first century. Nature Plants, 2: 1–8

Reid H, Song Y, Zhang Y, et al. (2018) Reducing Climate Risk and Poverty: Why China Needs Ecosystem-based Adaptation. International Institute for Environment and Development, London

Ricart S, Kirk N, Ribas A (2019) Ecosystem services and multifunctional agriculture: unravelling informal stakeholders' perceptions and water governance in three European irrigation systems. Environ Policy Gov, 29: 23–34

Ricke K, Caldeira K, Tavoni M, et al. (2018) Country-level social cost of carbon. Available via DIALOG https://socialsciences.nature.com/users/179738-katharine-ricke/posts/39211-countrylevel-social-cost-of-carbon. Accessed 20 Sep 2019

Ring I, Hansjürgens B, Elmqvist T, et al. (2010) Challenges in framing the economics of ecosystems and biodiversity: the TEEB initiative. Curr Opin Environ Sustain, 2: 15–26

Robbins L (1935) An Essay on the Nature and Significance of Economic Science. MacMillan and Company, London

Rodríguez-Labajos B, Martínez-Alier J (2013) The economics of ecosystems and biodiversity: recent instances for debate. Conserv Soc, 11: 326–342

Rowantree RA (1995) Toward ecosystem management: shifts in the core and the context of urban forest ecology. In: Bradley GA (ed) Urban forest landscapes-integrating multidisciplinary perspectives. University of Washington Press, Seattle, WA, pp 43–59

Rowantree RA, MacPherson EG, Nowak D (1994) The role of vegetation in urban ecosystems. In: MacPherson EG, Nowak DJ, Rowantree RA (eds).1994. Chicago's Urban Forest Ecosystem: Results of the Chicago Urban Forest Climate Project. General Technical Report NE-186. USDA Forestry Service. Northeast Forest Experiment Station. Washington DC, pp 1–3

Rowantree RA, Nowak D (1991) Quantifying the role of urban forests in removing atmospheric carbon dioxide. Journal of Arboriculture, 17(10): 269–275

Rowntree RA (1988) Ecology of the urban forest: introduction to part III. Landsc Urban Plan, 15(1–2): 1–10

Sagie H, Ramon U (2015) Using an agroecosystem services approach to assess tillage methods: a case study in the Shikma region. Land, 4: 938–956

Salafsky N, Salzer D, Stattersfield AJ (2008) A standard lexicon for biodiversity conservation: unified classifications of threats and actions. Conserv Biol, 22: 897–911

Sanderman J, Hengl T, Fiske GJ (2017) Soil carbon debt of 12, 000 years of human land use. Proc Natl Acad Sci, 114(36): 9575–9580

Sanders RA (1986) Urban vegetation impacts on the hydrology of Dayton, Ohio. Urban Ecol, 9 (3–4): 361–376

Sandhu H, Wratten S, Costanza R, et al. (2015) Significance and value of non-traded ecosystem services on farmland. Peer J, 3: e762

Sandhu HS, Crossman ND, Smith FP (2012) Ecosystem services and Australian agricultural enterprises. Ecol Econ, 74: 19–26

Sandhu HS, Wratten S, Porter J, et al. (2016) Mainstreaming ecosystem services into future farming. Solutions, 7: 40–47

Sandhu HS, Wratten SD, Cullen R (2008) The future of farming: the value of ecosystem services in conventional and organic arable land. An experimental approach. Ecol Econ, 64: 835–848

Sandhu HS, Wratten SD, Porter JR, et al. (2018) Biodiversityenhanced global agriculture. In: Ayton-Shenker D (ed) A new global agenda: priorities, practices, and pathways of the international community. Rowman & Littlefield, Lanham

Schlatter B, Trávníček J, Lernoud J, et al. (2020) Current statistics on organic agriculture worldwide: area, operators, and market. In: Willer H, Schlatter B, Trávníček J, et al. (eds) The World of Organic Agriculture. Statistics and Emerging Trends 2020. Research Institute of Organic Agriculture (FiBL), Frick, and IFOAM–Organics International, Bonn, pp 31–131

Schulte LA, Liebman M, Asbjornsen H, et al. (2006) Agroecosystem restoration through strategic integration of perennials. J Soil Water Conserv, 61: 164A–169A

Schumilas T (2014) Alternative Food Networks with Chinese Characteristics. University of Waterloo PhD thesis Waterloo, Ontario

Scott S, Si Z, Schumilas T, et al. (2014) Contradictions in state- and civil society-driven developments in China's ecological agriculture sector. Food Policy, 45(2): 158–166

Scott S, Si Z, Schumilas T, et al. (2018) Organic Food and Farming in China: Top-down and Bottom-up Ecological Initiatives. Routledge, London

Seufert V, Ramankutty N (2017) Many shades of gray—the context-dependent performance of organic agriculture. Sci Adv, 3

Seufert V, Ramankutty N, Foley JA (2012) Comparing the yields of organic and conventional agriculture. Nature, 485: 229–232

Shao M, Tang X, Zhang Y, et al. (2006) City clusters in China: air and surface water pollution. Front Ecol Environ, 4: 353–361

Sheil D (2018) Forests, atmospheric water and an uncertain future: the new biology of the global water cycle. Forest Ecosyst, 5: 236–257

Shepherd CJ, Jackson AJ (2013) Global fishmeal and fish-oil supply: inputs, outputs and markets. J Fish Biol, 83: 1046–1066

Shewaka SM, Mohamed N (2012) Green facades as a new sustainable approach towards climate change. Energy Procedia, 18: 507–520

Shi Y, Cheng C, Lei P, et al. (2011) Safe food, green food, good food: Chinese community supported agriculture and the rising middle class. Int J Agric Sustain, 9(4): 551–558

Si Z (2019) Shifting from industrial agriculture to diversified agroecological systems in China. The reclaiming diversity and citizenship series. Coventry University, Coventry

Si Z, Scott S (2016) The convergence of alternative food networks within "rural development" initiatives: the case of the new rural reconstruction movement in China. Local Environ, 21 (9): 1082–1099

Si Z, Scott S (2019) China's changing food system: top-down and bottom-up forces in food system transformations. Canad J Develop Stud. https://doi.org/10.1080/02255189.2019.1574005

Simpson JR (1998) Urban forest impacts on regional cooling and heating energy use: Sacramento case study. J Arboric, 24(4): 201–214

Sinha B, Hyung T (2008) Status and challenges in water infrastructure in Asia and the Pacific. The First regional workshop on the development of eco-efficient water infrastructure for socioeconomic development in Asia and the pacific region. Seoul, Republic of Korea, 10–12 Nov 2008

Smith P, Olesen JE (2010) Synergies between the mitigation of, and adaptation to, climate change in agriculture. J Agric Sci, 148: 543–552

Smith WH (1978) Urban vegetation and air quality. In: Proceedings of the National Urban Forestry Conference. ESF Publication 80-003, Syracuse, NY, pp 284–305

Somogyi S, Wang O, Charlebois S (2019) Mapping the value chain of imported shellfish in China. Mar Policy, 99: 69–75

Song X, Liu Y, Pettersen JB, et al. (2019) Life cycle assessment of recirculating aquaculture systems. A case of Atlantic salmon farming in China. J Ind Ecol, 23: 1077–1086

Song Y, Qi G , Zhang Y, et al. (2014) Farmer cooperatives in China: diverse pathways to sustainable rural development. Int J Agric Sustain, 12(2): 95–108

Souch CA, Souch C (1993) The effect of trees on summertime below canopy urban climates: a case study in Bloomington Indiana. J Arboric, 11(1): 1–12

SSSA (Soil Science Society of America) (n.d.) A new definition of soil. https://dl.sciencesocieties. org/publications/ csa/articles/62/10/20. Accessed 19 Jan 2020

State Council (2016) China's National Plan on Implementation of the 2030 Agenda for Sustainable Development. Available via DIALOG. http://www.greengrowthknowledge.org/nationaldocuments/chinas-national-plan-imp-lementation-2030-agenda-sustainabledevelopment

Stearns F, Montag T (1974) The Urban Ecosystem: a Holistic Approach. Dowden, Hutchinson & Ross, Stroudsburg, PA

Sternfeld E (2009) Organic food "made in China". EU-China Civil Society Review. 10-11 August 2009. pp 1–11. https://core.ac.uk/reader/10928838.Accessed 27 May 2021

Stonehouse P, Vander Borgh D (1995) Business Management Principles. Ontario Agricultural College access. University of Guelph, Guelph

Swanson D, Block R, Mousa SA (2012) Omega-3 fatty acids EPA and DHA: health benefits throughout life. Adv Nutr, 3: 1–7

Swinton SM, Lupi F, Robertson GP, et al. (2007) Ecosystem services and agriculture: cultivating agricultural ecosystems for diverse benefits. Ecol Econ, 64: 245–252

Tarasuk V, Mitchell A, Dachner N (2013) Household Food Insecurity in Canada. Available via DIALOG https://proof.utoronto.ca/food-insecurity/. Accessed 15 Mar 2020

Teletchea F, Fontaine P (2012) Levels of domestication in fish: implications for the sustainable future of aquaculture. Fish and Fisheries, 15: 181–195

Thrupp LA (2000) Linking agricultural biodiversity and food security: the valuable role of agrobiodiversity for sustainable agriculture. Int Af, 76: 265–281

Tilman D, Cassman KG, Matson PA, et al. (2002) Agricultural sustainability and intensive production practices. Nature, 418: 671–677

Tilman D, Hill J, Lehman C (2006) Carbonnegative biofuels from low-input high-diversity grassland biomass. Science, 314: 1598–1600

Tivy J (1990) Agricultural Ecology. Longman, New York, pp 288

Tivy J (2014) Agricultural Ecology. Routledge, Abingdon, pp 1

Todd-Brown KEO, Randerson JT, Hopkins F, et al. (2014) Changes in soil organic carbon storage predicted by earth system models during the 21st century. Biogeosciences, 11: 2341–2356

Tomich TP, Brodt S, Ferris H, et al. (2011) Agroecology: a review from a global-change perspective. Annu Rev Environ Resour 36: 193–222

Turner K, Lefler L, Freedman B (2004) Plant communities of selected urbanized areas of Halifax, Nova Scotia, Canada. Landsc Urban Plan, 71: 191–206

UFAW (1999) Animal sentience and the "five freedoms". https://www.ufaw.org.uk/about-ufaw/ufaw-and-animal-welfare. Accessed 10 Mar 2020

United Nations (2015) Sustainable Development Goals. Available via DIALOG. https://sustainabledevelopment. un.org/sdgs

United Nations (2015) Wetlands and Ecosystem Services. CBD press brief. Secretariat of the Convention on Biological Diversity, Montreal, QC. https://www.cbd.int/waters/doc/wwd2015/wwd-2015-press-briefs-en. pdf. Accessed 27 May 2021

United Nations (2018) World urbanization prospects: the 2018 revision (ST/ESA/SER.A/420). In: Department of Economic and Social Affairs, Population Division (2019). New York

United States Environmental Protection Agency (1994) Nonpoint source water pollution. USEPA Document #EPA

841-F-94-005. USEPA, Cincinnati, OH

United States Environmental Protection Agency (2009) Managing wet weather with green infrastructure. Available via DIALOG http://epa.gov/npdes/greeninfrastructure. Accessed 10 Jan 2020

Uphoff N, Thakur AK (2019) An agroecological strategy for adapting to climate change: the system of rice intensification (SRI). In: Sarkar A, Sensarma S, vanLoon G (eds) Sustainable Solutions for Food Security. Springer, Cambridge

van der Ploeg JD, Jingzhong Y, Schneider S (2012) Rural development through the construction of new, nested, markets: comparative perspectives from China , Brazil and the European Union. J Peasant Stud, 39(1): 133–173

Vanier Institute (2019) In focus 2019: food insecurity in Canada. Available via DIALOG https://vanierinstitute.ca/in-focus-2019-food-insecurity-in-canada/. Accessed 21 Mar 2020

Vitousek PM, Lubchenco J, Mooney H, et al. (1997) Human domination of earth ecosystems. Science, 277(5325): 494–499

Vogt G (2007) The origins of organic farming. In: Lockeretz W (ed) Organic Farming: an International History. CABI, Hershey, pp 9–29

Vörösmarty CJ, Hoekstra AY, Bunn SE, et al. (2015) Fresh water goes global. Science, 349: 478–479

Wang AM, Ge YX, Geng XY (2016a) Connotation and principles of ecological compensation in water source reserve areas based on the theory of externality. Chin J Population Resour Environ, 14: 189–196

Wang H, Dong Z, Xu Y, et al. (2016b) Eco-compensation for watershed services in China. Water Int, 41: 271–289

Wang J, Huang J, Zhang L, et al. (2010) Water governance and water use efficiency: the five principles of WUA management and performance in China. J Am Water Resour Assoc, 46: 665–685

Wang RY, Si Z, Ng CN, et al. (2015b) The transformation of trust in China's alternative food networks: disruption, reconstruction, and development. Ecol Soc, 20(2): 19

Wang SL (2005) Information technology: toward the way to the sustainable management of agroecosystem. Agriculture Network Inf, 8: 4–12

Wang SL (2007) Ecosystem services assessment and management. In: Lin WX (ed) Ecology. Science Press, Beijing

Wang SL, Caldwell CD, Kilyanek SL, et al. (2019) Using agroecology to stimulate the greening of agriculture in China: a reflection on 15 years of teaching and curriculum development. Int J Agric Sustain, 17: 298–311

Wang TF (2016) How to tell authentic bio-pesticides from fake ones. Farmers' Daily. http://www.yogeev.com/article/66984.html

Wang X, Cai D, Grant C, et al. (2015a) Factors controlling regional grain yield in China over the last 20 years. Agron Sustain Dev, 35(3): 1127–1138

Watson KA, Mayer AS, Reeves HW (2014) Groundwater availability as constrained by hydrogeology and environmental flows. Ground Water, 52: 225–238

WCED (1987) Our Common Future. Oxford University Press, New York

Weil RR, Brady NC (2017) The Nature and Property of Soils, 15th ed. Pearson, Columbus

Wezel A, Bellon S, Dore T, et al. (2009) Agroecology as a science, a movement and a practice. A review. Agron Sustain Dev, 29: 503–515

Wezel A, Goette J, Lagneaux E, et al. (2018) Agroecology in Europe: research, education, collective action networks, and alternative food systems. Sustainability, 10: 1214

WFP (World Food Program) (2016) 10 Facts about Nutrition in China. Available via DIALOG. https://www.wfp.org/stories/10-facts-about-nutrition-china

White R, Murray S, Rohweder M (2000) Pilot analysis of global ecosystems: grassland ecosystems. World Resources Institute. Washington, DC. http://pdf.wri.org/page_grasslands.pdf. Accessed 10 Oct 2019

Widdcombe RC, Carlisle B (1999) Geographic information and global positioning systems for tree management. J Arboric, 25(3): 175–178

Willer H, Schlatter B, Trávníček J, et al. (2020) In: Lernoud J (ed) The World of Organic Agriculture. Statistics and Emerging Trends 2020. Research Institute of Organic Agriculture (FiBL), Frick, and IFOAM-Organics International, Bonn

Williams C (1995) Healthy eating: clarifying advice about fruit and vegetables. Br Med J, 310: 1453–1455

Wilson EO (1988) The current state of biological diversity. In: Wilson EO (ed) Biodiversity. National Academic

Press, Washington, DC, pp 3–18

Wong C (2012) Guidance for the preparation of ESTR products-classifying threats to biodiversity. In: Canadian Biodiversity: Ecosystem Status and Trends 2010. Technical thematic report no. 2. Canadian Councils of Resource Ministers, Ottawa

Woodley A, Audette Y, Fraser T, et al. (2014) Nitrogen and phosphorus fertility management in organic field crop production. In: Martin RC, MacRae R (eds) Managing Energy, Nutrients and Pests in Organic Field Crops. CRC Press, Boca Raton, pp 59–106

World Health Organization (WHO) (1990) Diet, nutrition and the prevention of chronic disease. WHO. Technical Report Series 797, Geneva

Wratten SD, Sandhu H, Cullen R, et al. (2013) Ecosystem Services in Agricultural and Urban Landscapes. Wiley-Blackwell, Queensland, pp 200

Wright RJ (1984) Evaluation of crop rotation for control of Colorado potato beetle (Coleoptera: Chrysomelidae) in commercial potato fields on Long Island. Journal of Economic Entomology, 77: 1254–1259

Wu ZQ (1986) The Basic Agroecology. Fujian Science & Technology Press, Fuzhou

Xiao Q, McPherson E, Simpson J, et al. (1998) Rainfall interception by Sacramento's urban forest. J Arboric, 24(4): 235–244

Xinhua News (2018) The new age for strivers. Available via DIALOG. http://js.people.com.cn/n2/2018/0301/c360301-31295709.html

Yin S, Chen M, Chen Y, et al. (2016) Consumer trust in organic milk of different brands: the role of Chinese organic label. Br Food, J, 118(7): 1769–1782

Yiridoe EK, Weersink A (1997) A review and evaluation of agroecosystem health analysis: the role of economics. Agric Syst, 55(4): 601–626

Yousefi M, Khoramivafa M, Damghani AM (2017) Water footprint and carbon footprint of the energy consumption in sunflower agroecosystems. Environ Sci Pollut Res Int, 24 (24): 19827–19834

Ytrestøyl T, Aas TS, Åsgård T (2015) Utilization of feed resources in production of Atlantic salmon (*Salmo salar*) in Norway. Aquaculture, 448: 365–374

Zabel F, Delzeit R, Schneider JM, et al. (2019) Global impacts of future cropland expansion and intensification on agricultural markets and biodiversity. Nat Commun, 10: 2844

Zan CS, Fyles JW, Girouard P, et al. (2001) Carbon sequestration in perennial bioenergy, annual corn and uncultivated systems in southern Quebec. Agric Ecosyst Environ, 86: 135–144

Zedan H (2005) The role of the convention on biological diversity and its protocol on biosafety in fostering the conservation and sustainable use of the world's biological wealth for socioeconomic and sustainable development. J Ind Microbiol Biotechnol, 32: 496–501

Zhai HQ (1999) Introduction to agriculture. In: Toward 21st Century. Higher Education Press, Beijing

Zhang H (2019) Securing the "Rice bowl": China and Global Food Security. Palgrave Macmillan, Singapore

Zhang W, Lin J (2017) Research on the development strategy of community supported agriculture in China (*wo guo shequ zhichi nongye de fazhan celue yanjiu*). Acad J Zhongzhou, 10: 38–42

Zhong T, Si Z, Crush J, et al. (2019) Achieving urban food security through a hybrid public-private food 2 provisioning system: the case of Nanjing, China. Food Secur, 11(5): 1071–1086

Zhu WF (2011) Study on assessment of agroecosystem service & health and their managerial model in Fujian Province. Fujian Agriculture and Forestry University Master thesis, Fujian

Zhu WF, Wang SL, Caldwell CD (2010) Agroecosystem service and its managerial essence. Chin J Eco-Agric, 6: 889–896

Zhu WF, Wang SL, Caldwell CD (2012) Pathways of assessing agroecosystem health and agroecosystem management. Acta Ecol Sin, 32: 9–17

Zhu Y, Fen H, Wang Y, et al. (2000) Genetic diversity and disease control in rice. Nature, 406: 718–772

Zimmerer KS, de Haan S, Jones AD, et al. (2019) The biodiversity of food and agriculture (Agrobiodiversity) in the anthropocene: research advances and conceptual framework. Anthropocene, 25: 100192

Zolin MB, Cassion M, Mannino I (2017) Food Security, Food Safety and Pesticides: China and the EU Compared. Working paper. Foscari University of Venice, Venezia